CAMBRIDGE LIBRARY COLLECTION

Books of enduring scholarly value

Life Sciences

Until the nineteenth century, the various subjects now known as the life sciences were regarded either as arcane studies which had little impact on ordinary daily life, or as a genteel hobby for the leisured classes. The increasing academic rigour and systematisation brought to the study of botany, zoology and other disciplines, and their adoption in university curricula, are reflected in the books reissued in this series.

Sylva, Or, a Discourse of Forest Trees

John Evelyn (1620–1706), intellectual, diarist, gardener and founder member of the Royal Society, is best known for his *Diary*, the great journal of his life and times, encompassing a momentous period in British history. A lifelong collector of books, like his contemporary Pepys, Evelyn amassed over 4,000 items in his library. This work, originally published in 1664, was the first English-language treatise on forestry. Intended for the gentry, it aimed to encourage tree-planting after the ravages of the Civil War and to ensure a supply of timber for Britain's fast-developing navy. The first work sponsored officially by the Royal Society, it was an offshoot of Evelyn's unpublished manuscript *Elysium Britannicum*, a compendium of gardens and gardening. This is the 1908 two-volume reprint of the fourth edition, published in the year of Evelyn's death. Volume 2 covers practical aspects of forestry and the use of trees in landscaping.

Cambridge University Press has long been a pioneer in the reissuing of out-of-print titles from its own backlist, producing digital reprints of books that are still sought after by scholars and students but could not be reprinted economically using traditional technology. The Cambridge Library Collection extends this activity to a wider range of books which are still of importance to researchers and professionals, either for the source material they contain, or as landmarks in the history of their academic discipline.

Drawing from the world-renowned collections in the Cambridge University Library and other partner libraries, and guided by the advice of experts in each subject area, Cambridge University Press is using state-of-the-art scanning machines in its own Printing House to capture the content of each book selected for inclusion. The files are processed to give a consistently clear, crisp image, and the books finished to the high quality standard for which the Press is recognised around the world. The latest print-on-demand technology ensures that the books will remain available indefinitely, and that orders for single or multiple copies can quickly be supplied.

The Cambridge Library Collection brings back to life books of enduring scholarly value (including out-of-copyright works originally issued by other publishers) across a wide range of disciplines in the humanities and social sciences and in science and technology.

Sylva

Or, a Discourse of Forest Trees

*With an Essay on the Life
and Works of the Author*

VOLUME 2

JOHN EVELYN
EDITED BY JOHN NISBET

CAMBRIDGE
UNIVERSITY PRESS

CAMBRIDGE UNIVERSITY PRESS

Cambridge, New York, Melbourne, Madrid, Cape Town,
Singapore, São Paolo, Delhi, Mexico City

Published in the United States of America by Cambridge University Press, New York

www.cambridge.org
Information on this title: www.cambridge.org/9781108055277

© in this compilation Cambridge University Press 2013

This edition first published 1908
This digitally printed version 2013

ISBN 978-1-108-05527-7 Paperback

SYLVA: *OR A DISCOURSE*
OF FOREST TREES & THE
PROPAGATION OF TIMBER
V O L U M E   T W O

SYLVA

OR A DISCOURSE OF FOREST TREES : BY JOHN EVELYN F.R.S. *WITH AN ESSAY ON THE LIFE AND WORKS OF THE AUTHOR* BY JOHN NISBET D.Œc.

A REPRINT OF THE FOURTH
EDITION IN TWO VOLUMES
VOLUME TWO

LONDON : PUBLISHED BY ARTHUR
DOUBLEDAY & COMPANY LIMITED
AT 8 YORK BUILDINGS ADELPHI

DENDROLOGIA

THE THIRD BOOK

CHAPTER I.

Of Copp'ces.

1. *Sylva caedua* is (as Varro defines it) as well copp'ce to cut for fuel as for use of timber; and we have already shew'd how it is to be rais'd, both by sowing and planting. I shall only here add, that if in their first designation, they be so laid out, as to grow for several falls; they will both prove more profitable, and more delightful: More profitable, because of their annual succession; and more pleasant, because there will always remain some of them standing; and if they be so cast out, as that you leave straight and even intervals, of eighteen or twenty foot for grass, between spring-wood and spring-wood, securely fenc'd and preserv'd; the pastures will lie both warm, and prove of exceeding delight to the owner. These spaces are likewise useful and necessary for cartway, to fetch out the wood at every fall. There is not a more noble and worthy husbandry, than is this, which rejects no sort of ground nor situation, (tho' facing the east, is esteem'd best for both timber and under-wood) as we have abundantly shew'd; since even the most boggy places, may so be drein'd and cast, as to

AA

yield their increase by planting the dryer sorts upon the ridges and banks which you cast up, where they will thrive exceedingly : And then willow, sallow, alder, poplar, sycomor, black-cherry, &c. will shoot tolerably well on the lower and more uliginous ; with this caution, that for the first two years, they be kept diligently weeded and cleansed, which is as necessary as fencing, and guarding from cattle. Our ordinary copp'ces are chiefly upon hasle, or the birch ; but if amongst the other kinds, store of ash, (which I most prefer, a speedy and erect growth) chesnut, sallow, and sycomor, (at least one in four) were sprinkled in the planting, the profit would soon discover a differ-ence, and well recompence the industry. Others advise us to plant shoots of sallow, willow, alder, and all the swift-growing trees, being of seven years growth, sloping off both the ends towards the ground, to the length of a billet, and burying them a reason-able depth in the earth. This will cause them to put forth seven or eight branches, each of which will become a tree in a short time, especially if the soil be moist. The nearest distance for these plantations ought never to be less than five foot at first, since every felling renders them wider for the benefit of the timber, even to thirty and forty foot, in five or six fellings.

2. Though it be almost impossible for us to pre-scribe at what age it were best husbandry to fell copp'ces (as we at least call best husbandry) that is, for most and greatest gain ; since the markets, and the kinds of wood, and emergent uses do so much govern ; yet copp'ces are sometimes of a competent stature after eight or nine years from the acorn, and so every eight or ten years successively, will rise

better and better : But this had need be in extraordinary ground, otherwise you may do well to allow
them twelve or fifteen to fit them for the ax ; but
those of twenty years standing are better, and far
advance the price ; especially if oak, and ash, and
chesnut be the chief furniture ; and be sure you shall
lose nothing by this patience ; since all accidents consider'd, the profit arising from copp'ces so manag'd,
(be the ground almost never so poor) shall equal, if
not exceed what is usually made by the plough or
grazing. Some of our old clergy spring-woods heretofore have been let rest till twenty five or thirty
years, and have prov'd highly worth the attendance ;
for by that time, even a seminary of acorns, will
render a considerable advance, as I have already ex-
emplified in the Northamptonshire lady. And if
copp'ces were so divided, as that every year there
might be some fell'd, it were a continual, and a
present profit : Seventeen years growth affords a
tolerable fell ; supposing the copp'ce of seventeen
acres, one acre might be yearly fell'd for ever; and so
more, according to proportion ; but though the seldom
fall yields the more timber, yet the frequent makes
the under-wood the thicker ; therefore at ten or twelve
years growth (says Mr. Cook) in shallow ground, and
fourteen in deeper : If many timber-trees grow in
your copp'ces which are to be cut down, fell both
them, and the under-wood as near the ground as may
be ; but this is to be understood where the wood is
very thick ; otherwise, 'tis advisable to stock-up the
thinner, especially in great timber, and to set in the
holes, elm, cherry, poplar, sallow, service ; and so
these trees which are apt to grow from the running-
root thicken the wood exceedingly ; whilst the very

roots will pay for the grubbing, and yield you some feet of the best timber ; whereas being let stand, nothing would have grown : If the ground be a shallow soil, forbear filling the holes quite, but set some running-wood in the loosened earth, and the ends of the old roots being cut, will furnish the sides of the holes speedily : In thin copp'ces 'tis profitable to lay some boughs a-thwart, which will be rooted to advantage against next fall : All great rotten stubs among our under-woods should be extirpated, as making way for seedlings, and young roots to spring and run : The cutting, slanting, smooth, and close, is of great importance ; and frequent felling gives way and air to the subnascent seedlings, and the rest will make lusty shoots.

3. As to what numbers and scantlings you are to leave on every acre, the statutes are our general guides, at least the legal.   It is a very ordinary copp'ce, which will not afford three or four firsts, that is, bests ; fourteen seconds, twelve thirds, eight wavers, &c. according to which proportions, the sizes of young trees in copp'cing, are to succeed one another. By the statute of 35 Hen. 8. in copp'ces, or under-woods fell'd at twenty four years growth, there were to be left twelve standils, or stores of oak, upon each acre ; in defect of so many oaks, the same number of elms, ash, asp, or beech ; and they to be such, as are of likely trees for timber, and of such as have been spar'd at some former felling, unless there were none, in which case, they are to be then left, and so to continue without felling, till they are ten inch square within a yard of ground. Copp'ces above this growth fell'd, to leave twelve great oaks ; or in defect of them, other timber-trees (as above) and so

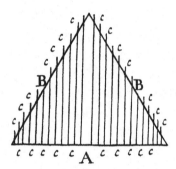

A, *Is the* Plain *of the* Basis, B B *the two sides of the* Triangle, *representing the sides of the* Mount. c c c c *the* Tops *of the* Trees, *shooting from the* Plain *and the sides.*

to be left for twenty years longer, and to be enclosed seven years.

4. In sum, you are to spare as many likely trees for timber, as with discretion you can. In the mean time, there are some who find it not so profitable, to permit so many timber-trees to stand in the heart of copp'ces ; but on the skirts, and near the edges, where their branches may freely spread, and have air, without dripping and annoying the subnascent crop : Nor should they be shread, which commonly makes them grow knotty. This is a note of the ingenious Mr. Nourse, as well as what he reports of a worthy gentleman in Gloucestershire, to demonstrate how one acre of copp'ce-wood on a plain, may contain as much wood as two acres on the side of an hill ; though that of the plain, as also the ground on the side of the hill, might seem both alike planted, and as thick in appearance.

For comparing the order in which trees usually grow on a plain, with those on a surface, they will appear standing exactly in such a figure : So that if the mountain be high and steep, one acre at the bottom may contain four times the quantity of wood, as an acre on the side of an hill, which is worth the consideration.

Now as to the felling (beginning at one side, that the carts may enter without detriment to what you leave standing,) the under-wood may be cut from January, at the latest, till mid-March or April ; or from mid-September, till near the end of November; so as all be avoided by Midsummer at the latest, and then fenced (where the rows and brush lie longer unbound or made up, you endanger the loss of a second-spring) and not to stay so long as usually they are a clearing, that the young, and the seedlings may suffer the least interruption ; And if the winter previous to your felling copp'ces, you preserve them well from cattle, it will recompence your care.

5. It is advis'd not to cut off the browse-wood of oaks in copp'ces, but to suffer it to fall off, as where trees stand very close, it usually does : I do not well comprehend why yet it should be spar'd so long.

6. When you espy a cluster of plants growing as it were all in a bunch, it shall suffice that you preserve the fairest sapling, cutting all the rest away. And if it chance to be a chesnut, service, or like profitable tree, clear it from the droppings and incumbrances of other trees, that it may thrive the better : Then, as you pass along, prune and trim-up all the young wavers, covering such roots as lie bare and expos'd, with fresh mould. There are some who direct the lopping of young oaks at a competent distance from the stem, and that while the wounds are healing this would advantage the under-wood ; but I cannot say it would be without prejudice to the timber.

7. Cut not above half a foot from the ground, nay the closer the better, and that to the south, slope-wise ; stripping up such as you spare from their extravagant branches, water-boughs, &c. that hinder the growth of others : Always remembring (before you so much as enter upon this work) to preserve sufficient plash-pole about the verge and bounds of the copp'ce for fence and security of what you leave; and for this, something less than a rod may suffice : Then raking your wood clear of spray, chips and all incumbrances, shut it up from the cattle, the longer the better.

8. By the statute, men were bound to enclose

copp'ces after felling, of, or under fourteen years growth, for four years : Those above fourteen years growth, to be sixteen years enclos'd ; and for woods in common, a fourth part to be shut up ; and at felling, the like proportion of great trees to be left, and seven years enclos'd : This was enlarg'd by 13 Eliz.   Your elder under-woods may be graz'd about July : But for a general rule, newly-weaned calves are the least noxious to newly-cut spring-woods, where there is abundance of grass ; and some say, colts of a year old ; but then the calves must be driven out at May at farthest, tho' the colts be permitted to stay a while longer : But of this, every man's experience will direct him ; and surely, the later you admit beasts to graze, the better.   For the measure of fuel, these proportions were to be observ'd.

9. Statutable billet should hold three foot in length, and seven inch and half compass ; ten or fourteen as they are counted for one, two, or three, &c.   A stack of wood (which is the boughs and offal of the trees to be converted to char-coal) is four yards long, three foot and half high (in some places but a yard) and as much over : In other places, the cord is four foot in height, and four foot over; or (to speak more geometrically) a solid made up of three dimensions, four foot high, four foot broad, and eight foot long ; the content 128 cubique feet.   Faggots, ought to be a full yard in length, and two foot in circumference, made round, and not flat; for so they contain less fuel, though equal in the bulk appearing.   But of these particulars, when we come to speak expresly of fuel.

10. In the mean time it were to be wish'd, that some approved experiments were sedulously try'd (with the advice of skilful and ingenious physicians)

for the making of beer without hops ; as possibly
with the white *marrubium* (a plant of singular virtue)
or with dry'd heath-tops, (*viz.* that sort which bears
no berries) or the like, far more wholesome, and less
bitter then either, tamarisk, *carduus*, or broom, which
divers have essay'd ; it might prove a means to save
a world of fuel, and in divers places young timber,
and copp'ce-wood, which is yearly spent for poles ;
especially in countries where wood is very precious.

Note, that the wood-land measure by statute, is
computed after eighteen foot the perch.

CHAPTER II.

Of Pruning.

There could nothing certainly be more necessary,
in order to pruning, than the knowledge of the course
and nature of the sap ; which not being as yet so
universally agreed on (after innumerable tryals and
experiments) leads our arborators into many errors
and mistakes: I have in this Forest Work occasionally
recited the various opinions of several, leaving them
to the determination of the learned and judicious, as
a considerable part of natural philosophy ; Dr. Grew,
Malphigius, De la Quinteny, and what is found dispers'd in the *Philos. Transactions* by our plant anatom-
ists ; without charging this chapter with repetitions:
And the same I have done likewise as to astrological
observations, positions of the stars and planetary
configurations, exhalations and dominant power ;
though in compliance to custom, I now and then

forbear to abdicate our country planter's goddess ;
contenting my self with the wholsomeness of the air
we breathe in, and the goodness of the soil : I shall
therefore in the first place speak of the manual operation of pruning, and other instructions as they afterwards occur :

1. *Putatio ;* pruning I call all purgation of trees in
general, from what is superfluous : The Ancients
found such benefit in pruning, that they feigned a
Goddess presided over it, as Arnobius tells us : And
in truth, it is in the discreet performance of this work,
that the improvement of our timber and woods does
as much consist as in any thing whatsoever.   A skilful planter should therefore be early at this work :
Shall old Gratius give you reason and direction ?
And his interpreter thus in English ?

[1] Twigs of themselves never rise straight and high,
And under-woods are bow'd as first they shoot.
Then prune the boughs ; and suckers from the root
Discharge.   The leavy wood fond pity tires.
After, when with tall rods the tree aspires,
And the round staves to heaven advance their twigs,
Pluck all the buds, and strip off all the sprigs ;
These issues vent what moisture still abound,
And the veins unemploy'd grow hard and sound.
                      Wase.

2. For 'tis a misery to see how our fairest trees are
defac'd, and mangled by unskilful wood-men, and

[1] Nunquam sponte sua procerus ad aera termes
Exiit, inque ipsa curvantur stirpe genestæ.
Ergo age luxuriam primo fœtusque nocentes
Detrahe. Frondosas gravat indulgentia silvas.
Post ubi proceris generosa stirpibus arbor
Se dederit, teretesque ferent ad sidera virgae,
Stringe notas circum, & gemmantes exige versus.
His, si quis vitium nociturus sufficit humor,
Visceribus fluit, & venas durabit inertes.
              Gra. Fal. *Cyneget.*

mischievous bordurers, who go always armed with
short hand-bills, hacking and chopping off all that
comes in their way; by which our trees are made full
of knots, stubs, boils, cankers, and deform'd bunch-
es, to their utter destruction : Good husbands should
be asham'd of it ; tho' I would have no wood-man
pretend to be without all his necessary furniture,
when he goes about this work ; which I (once for
all) reckon to be the hand-bill, hatchet, hook, hand-
saw, an excellent pruning-knife, broad chizel and
mallet, all made of the best steel and kept sharp ; and
thus he is provided for greater, or more gentle exe-
cutions, purgations, recisions, and coercions; and it is
of main concern, that the proper and effectual tool be
applied to every work, since heavy and rude instruments do but mangle and bruise tender plants; and if
they be too small, they cannot make clear and even
work upon great arm and branches: The knife is for
twigs and spray ; the chizel for larger armes, and such
amputations as the ax and bill cannot well operate
upon.   As much to be reprehended are those who
either begin this work at unseasonable times, or so
maim the poor branches, that either out of laziness,
or want of skill, they leave most of them stubs, and
instead of cutting the arms and branches close to the
bole, hack them off a foot or two from the body of
the tree, by which means they become hollow and
rotten, and are as so many conduits to receive the
rain and the weather, which conveys the wet to the
very *matrix* and heart, deforming the whole tree with
many ugly botches, which shorten its life, and utterly
mars the timber : I know Sir H. Platt tells us, the
elm should be so lopp'd, but he says it not of his
own experience as I do.   And here it is that I am

(once for all) to warn our disorderly husband-men from coveting to let their lops grow to an extraordinary size, before they take them off, as conceiving it furnishes them with the more wood for the fire ; not considering how such gashly wounds mortally affect the whole body of the tree, or at least does so decay their vigour, that they hereby lose more in one year, than the lop amounts to, should they pare them off sooner, and when the scars might be cover'd: In the mean while, that young oaks prosper much in growth, by timely pruning, the industrious Mr. Cooke observes ; whereas some other trees, as the horn-beam, &c. though they will bear considerable lops, when there's only the shell of the tree standing, yet it is much to its detriment ; especially to the ash, which if once it comes to take wet by this means, rarely produces more lop to any purpose ; above all, if it decay in the middle, when 'tis fitter for the chimney, than to stand and cumber the ground : The same may be pronounc'd of most trees, which would not perhaps become dotards in many ages, but for this covetous barbarity, and unskilful handling.

3. By this animadversion alone it were easy for an ingenious man to understand how trees are to be govern'd ; which is in a word, by sparing great lops, cutting clean, smooth, and close, making the stroke upward, and with a sharp bill, so as the weight of an untractable bough do not splice, and carry the bark with it, which is both dangerous and unsightly ; the oak will suffer it self to be made a pollard, that is, to have its head quite cut off, and it may be good for mast, if not too much prun'd, but not for timber : But the elm so treated, will perish to the foot, and certainly become hollow at last, if it 'scape with life.

4. The proper season for this work, is for old trees earlier, for young later, as a little after the change in January or February, some say in December, the wind in a gentle quarter :

[1] Then shave their locks, and cut their branchy tress,
 Severely now, luxuriant boughs repress.

But this ought not to be too much in young fruit-trees, after they once come to form a handsome head; in which period you should but once pare them over about March, to cover the stock the sooner, if the tree be very choice : To the aged, this is plainly a renewing of their youth, and an extraordinary refreshment, if taken in time, and that their arms be not suffer'd to grow too great and large; in which case, the member must not be amputated too near the body, but at some distance — *ne pars sincera trahatur :* And remember to cut smooth, and sloping upwards if upright boughs, otherwise downwards; and be sure to emplaster great wounds to keep out the wet, and hasten the covering of the bark: besides, for interlucation, exuberant branches, &c. *spissae nemorum comae,* where the boughs grow too thick, and are cumbersome, to let in the sun and air; this is of great importance; and so is the sedulous taking away of suckers, waterboughs, fretters, &c. And for the benefit of tall timber, the due stripping up the branches, and rubbing off the buds to the heights you require: Yet some do totally forbear the oak, especially if aged, observing that they much exceed in growth such as are prun'd; and in truth such trees as we would leave for shade and

---

[1] .... Tunc stringe comas, tunc brachia tonde :
 .... Tunc denique dura
 Exerce imperia, & ramos compesce fluentes.
                                          Georg. 2.

ornament, should be seldom cut ; but the brouse-wood cherish'd and preserv'd as low towards the ground as may be, for a more venerable and solemn shade : And therefore I did much prefer the walk of elms at St. James's Park, as it lately grew branchy, intermingling their reverend tresses, before the present trimming them up so high ; especially since, I fear, the remedy comes too late to save their decay, (could it have been avoided) if the amputations of such overgrown parts as have been cut off, should not rather accelerate it, by exposing their large and many wounds to the injuries of the weather, which will indanger the rotting of them, beyond all that can be apply'd by tar, or otherwise to protect them : I do rather conceive their infirmities to proceed from what has not long since been abated of their large spreading branches, to accommodate with the Mall ; as any one may conjecture by the great impression which the wet has already made in those incurable scars, that being now multiplied, must needs the sooner impair them ; the roots having likewise infinitely suffer'd, by many disturbances about them.   In all events this walk might have enjoy'd its goodly canopy with all their branchy furniture for some ages to come ; since 'tis hardly one, that first they were planted : But this defect is providently and nobly supply'd, by their successors of the lime-trees, which will sooner accomplish their perfection, by taking away the chesnut trees, which will else do them prejudice.

But it is now (and never till now) that those walks and ranks of trees, and other royal amenities, are sure to prosper, whilst they are entirely under the care and culture of the most industrious and knowing Mr. Wise, (to whom, and to his partner Mr. London) I not

only acknowledge my self particularly oblig'd ; but
the whole nation for what they have contributed to
the sweetest, useful, and most innocent diversions of
life, gardens and plantations.

One should be cautious in heading timber-trees,
especially the pithy ; unless where they grow very
crooked, in which case abate the head with an upward
sloop, and cherish a leading shoot : The beech is very
tender of its head.

It is by the discreet leaving the side-boughs in convenient places, sparing the smaller, and taking away
the bigger, that you may advance a tree to what
determin'd height you desire : Thus, bring up the
leader, and when you would have that spread and
break out, cut off all the side-boughs, and especially
at Midsummer, if you espy them breaking out.
Young trees may every year be prun'd, and as they
grow older at longer intervals, as at three, five, seven,
or sooner, that the wounds may recover, and nothing
be deformed.

Ever-greens do not well support to be decapitated ;
side-boughs they freely spare in April, and during the
Spring ; and if you cut at first two or three inches
from the body, and the next Spring after, close to the
stem, covering it with wax, or well temper'd clay, the
most tender may suffer such amputations without
prejudice.

Note that the side and collateral branches of the
fir, cut, or broken off, spring no more ; and though
the tops sometimes do, yet they never prosper to
beautiful and erect heads, in which consists the grace
of that beautiful tree.

Another caution is, that you be sure to cut off such
tender branches to the quick, which you find have

been cropt by goats, or any other cattle, who leave a
drivel where they bite ; which not only infects the
branches, but sometimes indanger the whole ; the
reason is, for that the natural sap's recourse to the
stem, communicates the venom to all the rest, as the
whole mass and habit of animal blood is by a gangreen,
or venereal taint.

5. Divers other precepts of this nature I could here
enumerate, had not the great experience, faithful and
accurate description how this necessary work is to be
perform'd, set down by our countryman honest Lawson
(*Orchards*, cap. 11) prevented all that the most inquisitive can suggest: The particulars are so ingenious,
and highly material, that you will not be displeas'd
to read them in his own style and character.

All ages (saith he) by rules and experience do
consent to a pruning and lopping of trees : Yet have
not any that I know described unto us (except in
dark and general words) what, or which are those
superfluous boughs which we must take away ; and
that is the most chief, and most needful point to be
known in lopping.   And we may well assure our
selves (as in all other arts, so in this) there is a vant-
age and dexterity by skill ; an habit by practice out
of experience, in the performance hereof, for the
profit of mankind : Yet do I not know (let me speak
it with patience of our cunning arborists) any thing
within the compass of human affairs so necessary and
so little regarded ; not only in orchards, but also in
all other timber-trees, where or whatsoever.

Now to our purpose :

How many forests and woods, wherein you shall
have for one lively thriving tree, four (nay sometimes
twenty four) evil thriving, rotten and dying trees,

even whiles they live ; and instead of trees, thousands
of bushes and shrubs ! What rottenness ! what
hollowness ! what dead arms ! withered tops ! cur-
tail'd trunks ! what loads of moss ! drouping boughs,
and dying branches shall you see every where ! and
those that in this sort are in a manner all unprofitable
boughs, canker'd arms, crooked, little and short boals.
What an infinite number of bushes, shrubs, and
skrags of hasels, thorns, and other unprofitable wood,
which might be brought by dressing to become great
and goodly trees ! Consider now the cause.

The lesser wood hath been spoil'd with careless,
unskilful, and untimely stowing ; and much also of
the great wood. The greater trees at the first rising
have fill'd and overladen themselves with a number
of wastful boughs and suckers, which have not only
drawn the sap from the boal, but also have made it
knotty, and themselves, and the boal mossie, for want
of dressing ; whereas, if in the prime of growth, they
had been taken away close, all but one top, and
clean by the bulk, the strength of all the sap should
have gone to the bulk, and so he would have recov-
ered and covered his knots, and have put forth a fair,
long and streight body, for timber profitable, huge
great of bulk, and of infinite last.

If all timber trees were such, (will some say) how
should we have crooked wood for wheels, coorbs, &c.

Answ. Dress all you can, and there will be enough
crooked for those uses.

More than this, in most places they grow so thick,
that neither themselves, nor earth, nor any thing
under or near them can thrive ; nor sun, nor rain, nor
air can do them, nor any thing near, or under them,
any profit or comfort.

I see a number of hags, where out of one root you shall see three or four (nay more, such is men's unskilful greediness, who desiring many, have none good) pretty oaks, or ashes streight and tall ; because the root at the first shoot gives sap amain : But if one onely of them might be suffered to grow, and that well and cleanly prun'd, all to his very top, what a tree should we have in time ? And we see by those roots continually and plentifully springing, notwithstanding so deadly wounded, what a commodity should arise to the owner, and the Commonwealth, if wood were cherished and orderly dressed. The waste boughs closely and skilfully taken away, would give us store of fences and fuel ; and the bulk of the tree in time would grow of huge length and bigness: But here (methinks) I hear an unskilful arborist say, that trees have their several forms, even by nature ; the pear, the holly, the aspe, &c. grow long in bulk, with few and little armes. The oak by nature broad, and such like. All this, I grant : But grant me also, that there is a profitable end and use of every tree, from which if it decline (though by nature) yet man by art may (nay must) correct it. Now other end of trees I could never learn, than good timber, fruit much and good, and pleasure : uses physical hinder nothing a good form.

Neither let any man ever so much as think, that it is unprofitable, much less unpossible, to reform any tree of what kind soever : for (believe me) I have tried it : I can bring any tree (beginning betime) to any form. The pear, and holly may be made to spread, and the oak to close.

Thus far the good man out of his eight and forty years experience concerning timber-trees : He des-

cends then to the orchards ; which because it may
likewise be acceptable to our industrious planter, I
thus contract.

6. Such as stand for fruits should be parted from
within two foot (or thereabouts) of the earth ; so
high, as to give liberty to dress the root, and no
higher ; because of exhausting the sap that should
seed his fruit : For the boal will be first, and best
served and fed, being next to the root, and of greatest
substance. These should be parted into two, three,
or four arms, as your graffs yield twigs ; and every
arm into two, or more branches, every branch into
his several cyons ; still spreading by equal degrees ;
so as his lowest spray be hardly without the reach of
a man's hand, and his highest not past two yards
higher : That no twig (especially in the middest)
touch his fellow ; let him spread as far as his list,
without any master-bough, or top, equally ; and
when any fall lower than his fellows (as they will
with weight of fruit) ease him the next spring of his
superfluous twigs, and he will rise : When any mount
above the rest, top him with a nip between your
fingers, or with a knife : Thus reform any cyon ; and
as your tree grows in stature and strength, so let him
rise with his tops, but slowly, and easily, especially in
the midst, and equally in breadth also ; following
him upward, with lopping his undergrowth, and
water-boughs, keeping the same distance of two yards,
not above three, in any wise, betwixt the lowest and
highest twigs.

1. Thus shall you have handsome, clear, healthful,
great and lasting trees.

2. Thus will they grow safe from winds, yet the
top spreading.

3. Thus shall they bear much fruit; I dare say, one as much as five of our common trees, all his branches loaden.

4. Thus shall your boal being low, defraud the branches but little of their sap.

5. Thus shall your trees be easie to dress, and as easie to gather the fruit from, without bruising the cyons, &c.

6. The fittest time of the moon for the pruning is (as of graffing) when the sap is ready to stir (not proudly stirring) and so to cover the wound; and here, for the time of day, we may take Columella, *frondem medio die arborator ne caedito*, l. 11. Old trees would be prun'd before young plants: And note, that wheresoever you take any thing away, the sap the next Summer will be putting; be sure therefore when he puts to bud in any unfit place, you rub it off with your finger; and if this be done for three or four years still at Midsummer, it will at last wholly clear the side-boughs, and exalt the growth of the stem exceedingly; and this is of good use for elms, and such trees as are continually putting forth where they have been prun'd: Thus begin timely with your trees, and you may bring them to what form you please. If you desire any tree should be taller, let him break, or divide higher: This for young trees: The old are reformed by curing of their diseases, of which we have already discours'd. There is this only to be considered, in reference to foresters, out of what he has spoken concerning fruit-trees; that (as has been touch'd) where trees are planted for shadow, and meer ornament, as in walks and avenues, the brousewood (as they call it) should most of it be cherished; whereas in fruit, and timber-trees (oak excepted) it is

best to free them of it : As for pollards (to which
I am no great friend, because it makes so many scrags
and dwarfes of many trees which would else be good
timber, endangering them with drips and the like
injuries) they should not be headed above once in ten
or twelve years, at the beginning of the Spring, or
end of the fall.   And note, that all copp'cing and
cutting close, invigorates the roots and the stem of
whatsoever grows weak and untimely ; but you must
then take care it be not overgrown with weeds or
grass : Nothing (says my Lord Bacon *exp*. 586. and
truly) causes trees to last so long, as the frequent
cutting; every such diminution being a re-invigor-
ation of the plant's juice, so that it neither goes
too far, nor rises too faintly, as when 'tis not timely
refresh'd with this remedy ; and therefore we see,
that the most ancient trees in church-yards, and
about old buildings, are either pollards or dotards,
seldom arising to their full altitude.   'Tis true (as
Mr. Nourse observes) that elm and oak frequently
pollarded and cut, hindering their mounting, increases
the bulk and circumference, and makes a show
of substance; when all the while 'tis but a hollow
trunk, fill'd with its own corruption, spending the
genuine moisture which should go to the growth of
the arms and head, and interior substance of useful
timber.

7. For the improvement of the speedy growth of
trees, there is not a more excellent thing than the
frequent rubbing of the boal or stem, with some
piece of hair-cloth, or ruder stuff, at the beginning
of Spring: Some I have known done with seals-skin;
the more rugged bark with a piece of coat of mail,
which is made of small wyres : This done, when the

body of the trees are wet, as after a soaking rain ; yet so, as not to excorticate, or gall the tree, has exceedingly accelerated its growth, (I am assured, to a wonderful and incredible improvement) by opening the pores, freeing them of [1] moss, and killing the worm.

8. Lastly, frondation, or the taking off some of the luxuriant branches and sprays of such trees, especially whose leaves are profitable for cattle (whereof already) is a kind of pruning: And so is the scarrifying and cross hatching of some fruit-bearers, and others, to abate that $\phi\upsilon\lambda\lambda o\mu\alpha\nu\acute{\iota}\alpha$ which spends all the juice in the leaves, to the prejudice of the rest of the parts.

But after all this, let us hear what the learned and experienc'd Esq ; Brotherton has observ'd upon this article of pruning, and particularly of the taking of the top ; that those trees which were so us'd, some years before the severe frost of 1684, died: Those not so prun'd, escap'd : And of other trees, (having but a small head left) the rest of the boughs cleared ; the tops flourish'd, and the loose branches shread, perish'd, and the unprun'd escap'd : Moreover, when the like pruning has been try'd on trees 20 foot high ; the difference of the increase was visible the following Summer ; but within 7 or 8 years time, the difference was exceeding great, and even prodigious, both in bark and branch, beyond those trees that had been prun'd.

9. This, and the like, belonging to the care of the wood-ward, will mind him of his continual duty ; which is to walk about, and survey his young plantations daily ; and to see that all gaps be immediately

---

[1] See cap. 7. book 2.

stopt ; trespassing cattle impounded ; and (where they are infested) the deer chased out, &c. It is most certain, that trees preserv'd and govern'd by this discipline, and according to the rules mention'd, would increase the beauty of forests, and value of timber, more in ten or twelve years, than all other imaginable plantations (accompanied with our usual neglect) can do in forty or fifty.

10. To conclude, in the time of this work should our ingenious arborator frequently incorporate, mingle, and unite the arms and branches of some young and flexible trees which grow in consort, and near to one another ; by entring them into their mutual barks with a convenient incision : This, especially, about fields and hedge-rows, for fence and ornament. Dr. Plot mentions some that do naturally, or rather indeed accidentally mingle thus ; nay, and so imbrace and coalesce, as if they issu'd out of the bowels of one another : Such are the two beeches in the way from Oxford to Reading at Cain-End ; the bodies of which trees springing from different roots, after they have ascended parallel to the top, strangely unite together a great height from the ground, a transverse piece of timber entring at each end the bodies of the trees, and growing jointly with them : The same is seen in sycomores at New-Colledge gardens : I my self have woven young ash-poles into twists of three and four braids, like womens hair, when they make it up to fillet it under their coifes, which have strangely incorporated and grown together without separation ; but these are rather for curiosity, than of advantage for timber.

Trees will likewise grow frequently out of the boal of the other, and some roots will penetrate

through the whole length of the trunk, till fastening
in the very Earth, they burst the including tree, as it
has happened in willows, where an ash-tree has sprung
likely from some key or seed dropt upon the rotten
head of it : But this accident not so properly pertain-
ing to this chapter, I conclude with recommending
the bowing and bending of young timber-trees, especially oak and ash, into various flexures, curbs, and
postures, oblig'd to ply themselves into different
modes, which may be done by humbling and binding
them down with tough bands and withs, or hooks
rather, cut skrew-wise, or slightly hagled and indented with a knife, and so skrewed into the ground, or
hanging of weighty stones to the tops, or branches,
till the tenor of the sap, and custom of being
so constrain'd, did render them apt to grow so
of themselves, without power of redressing : This
course would wonderfully accommodate materials
for knee-timber and shipping, the wheel-wright,
and other uses ; conform it to their moulds, and
save infinite labour, and abbreviate the work of hewing and waste.

.........*adeo in teneris consuescere multum est.*

and the poet, it seems, knew it well, and for what
purposes,

> [1] When in the woods with mighty force they bow
> The elm, and shape it to a crooked plow.

so as it even half-made it to their hands.

---

[1] Continuò in silvis magna vi flexa domatur
In burim, & curvi formam accipit ulmus aratri.
Geo. I.

CHAPTER III.

Of the Age, Stature, and Felling of Trees.

1. The age of trees, except of the coniferous, (for
the most part known by the degrees of their tapering
branches) is vulgarly reckon'd by the number of solar
revolutions, or circles ; the former bark being digested
and compacted into lignous and woody substance,
which is annually invested by a succeeding bark ;
which yet in some is not finish'd so soon as in other
trees, as we find in the oak, elm, pine, plum-trees,
&c. which exceed one another in growth, however
coaequal in years ; But of this hereafter. In the
mean time, it is not till a tree is arriv'd to his perfect
age and full vigor, that the lord of the forest should
consult or determine concerning a felling. For there
is certainly in trees (as in all things else) a time of
increment, or growth ; a *status* or season when they
are at best, (which is also that of felling) and a
decrement or period when they decay. To the first
of these they proceed with more or less velocity, as
they consist of more strict and compacted particles,
or are of a slighter and more laxed contexture ; by
which they receive a speedier or slower defluxion of
aliment. This is apparent in box, and willow ; the
one of a harder, the other of a more tender substance:
But as they proceed, so they likewise continue. By
the state of trees I would signifie their utmost effort,
growth, and maturity, which are all of them different
as to time, and kind ; yet do not I intend by this any
period or instant in which they do not continually

either improve or decay, (the end of one being still
the beginning of the other) but farther than which
their natures do not extend ; but immediately (though
to our senses imperceptibly) through some infirmity
(to which all things sublunary be obnoxious) dwindle
and impair, either through age, defect of nourishment,
by sickness and decay of principal parts ; but especially and more inevitably, when violently invaded by
mortal and incurable infirmities, or by what other
extinction of their vegetative heat, substraction, or
obstruction of air and moisture ; which making all
motions whatsoever to cease and determine, is the
cause of their final destruction.

2. Our honest countreyman, to whose experience
we have been obliged for something I have lately
animadverted concerning the pruning of trees, does
in another chapter of the same treatise, speak of the
age of trees. The discourse is both learned, rational,
and full of encouragement : For he does not scruple
to affirm, that even some fruit-trees may possibly
arrive to a thousand years of age ; and if so fruit-
trees, whose continual bearing does so much impair
and shorten their lives, as we see it does their form
and beauty ; how much longer might we reasonably
imagine some hardy and slow-growing forest-trees
may probably last ? I remember Pliny tells us of
some oaks growing in his time in the [1] Hercynian
forest, which were thought co-evous with the world
it self ; their roots had even raised mountains, and
where they encounter'd swell'd into goodly arches,
like the gates of a city : But our more modern
author's calculation for fruit-trees (I suppose he means

[1] Hercyniæ silvæ roborum vastitas intacta aevis, & congenita mundo, prope
immortali sorte miracula excedit. Plin. l. 16. c. 2.

pears, apples, &c.) his allowance is three hundred
years for growth, as much for their stand (as he terms
it,) and three hundred for their decay, which does in
the total amount to no less than nine hundred years.
This conjecture is deduc'd from apple-trees growing
in his orchard, which having known for forty years,
and upon diligent enquiry of sundry aged persons of
eighty years and more, who remembred them trees
all their time, he finds by comparing their growth
with others of that kind, to be far short in bigness
and perfection, (*viz.* by more than two parts of three)
yea albeit those other trees have been much hindred
in their stature, through ill government and mis-
ordering : And this to me seems not at all extravagant, since I find mention of a pear-tree near Ross in
Herefordshire, which being of no less than 18 foot in
circumference, and yielding seven hogsheads of cider
yearly, must needs have been of very long standing
and age, tho' perhaps not so near Methusalem's.

3. To establish this, he assembles many arguments
from the age of animals, whose state and decay double
the time of their increase by the same proportion :
If then (saith he) those frail creatures, whose bodies
are nothing (in a manner) but a tender rottenness,
may live to that age ; I see not but a tree of a solid
substance, not damnified by heat or cold, capable of,
and subject to any kind of ordering or dressing, feeding
naturally, and from the beginning disburthen'd of all
superfluities, eased of, and of his own accord avoiding
the causes that may annoy him, should double the
life of other creatures by very many years. He proceeds, What else are trees in comparison with the
earth, but as hairs to the body of man ? And it is
certain, that (without some distemper, or forcible

cause) the hairs dure with the body, and are esteem'd
excrements but from their superfluous growth : So
as he resolves upon good reason, that fruit-trees well
ordered may live a thousand years, and bear fruit ;
and the longer the more, the greater, and the better ;
(for which an instance also in Dr. Beal's *Herefordshire
Orchards*, pag. 21, 22.) because his vigour is proud
and stronger, when his years are many. Thus you
shall see old trees put forth their buds and blossoms
both sooner, and more plentifully than young trees
by much ; and I sensibly perceive (saith he) my
young trees to enlarge their fruit as they grow greater,
&c. and if fruit-trees continue to this age, how many
ages is it to be supposed strong and huge timber-trees
will last ? whose massy bodies require the years of
divers Methuselahs, before they determine their days ;
whose sap is strong and bitter ; whose bark is hard
and thick, and their substance solid and stiff ; all
which are defences of health and long life. Their
strength withstands all forcible winds ; their sap of
that quality is not subject to worms and tainting ;
their bark receives seldom or never by casualty any
wound ; and not only so, but they are free from
removals, which are the death of millions of trees ;
whereas the fruit-tree (in comparison) is little, and
frequently blow down ; his sap sweet, easily and soon
tainted ; his bark tender, and soon wounded ; and
himself used by man as man uses himself ; that is,
either unskilfully, or carelesly. Thus he. But
Vossius *de theolog. gent.* lib. 5. c. 5. gives too little age
to ashes, when he speaks but of one hundred years,
(in which, as in the rest, he seems to agree with my
Lord Bacon, *hist. vitae & mort. artic.* 1.) and to the
Medica, pyrus, prunus, cornus but sixty ; he had as good

have held his peace : Even rosemary has lasted amongst us a hundred years.

4. I might to this add much more, and truly with sufficient probability, that the age of timber-trees, especially of such as be of a compact, resinous, or balsamical nature, (for of this kind are the yew, box, horn-beam, white-thorn, oak, walnut, cedar, juniper, &c.) are capable of very long duration and continuance : Those of largest roots (a sign of age) longer liv'd than the shorter ; the dry than the wet ; and the gummy, than the watery ; the sterile, than the fruitful: For not to conclude from Pliny's [1] Hercynian oaks, or the turpentine tree of Idumaea, (which Josephus ranks also with the creation :) I mention'd a cypress yet remaining somewhere in Persia near an old sepulchre, whose stem is as large as five men can encompass, the boughs extending fifteen paces every way ; this must needs be a very old tree, believ'd by my author little less than 2500 years of age. Of such another, Dr. Spon in his voyage into Greece speaks, which by its spreading seems to be of the savine-kind: And in truth, as to the age and duration, cypress, cedar, box, ebony, Brasil, and other exceeding hard and compact (with some resinous) woods, growing chiefly in both East and West-Indies, must needs be of wonderful age. The particulars were too long to recount. The old *platanus* set by Agamemnon, mention'd by Theophrastus, and the Herculean oaks ; the laurel near Hippocrene, the Vatican Ilex, the vine which was grown to that bulk and woodiness, as to make a statue of Jupiter and columns in Juno's temple ; and at present 'tis found that the great doors

---

[1] *Silvarum, Hercynia dierum sexaginta iter occupans, ut major aliis, ita et notior.* Pomp. Mela. l. 3. c. 3.

of the cathedral at Ravena is made of such vine-tree
planks ; some of which are 12 foot long, 14 and 15
inches broad ; the whole soil of that country producing vines of prodigious growth ; and such another in
Margiana is spoken of by Strabo, that was twelve
foot in circumference : Pliny mentions one of six
hundred years old in his time ; and at Ecoan the late
Duke of Montmorancy's house, is a table of a very
large dimension, made of the like plant : And that
which renders it the more strange, is, that a tree
growing in such a wreath'd and twisted manner, rather
like a rope than timber, and needing the support of
others, should arrive to such a bulk, and firm consistence ; but so it is, and Olearius affirms, that he
found many vines near the Caspian Sea, whose trunks
were as big about as a man.   And the old Lotus
trees, recorded by Valerius Maximus, and the *Quercus
Mariana*, celebrated by the prince of orators : Pliny's
huge *larix*, and what grew in the Fortunate Islands,
with that enormous tree Scaliger reports was growing
in the Troglodytic India, &c. were famous for their
age : St. Hierom affirms he saw the sycomor that
Zaccheus climb'd up, to behold our Lord ride in
triumph to Jerusalem : But that's nothing for age to
the olive, under which our Blessed Saviour agoniz'd,
still remaining (as they say) in the garden to which
he us'd to resort.   At the same rate, Surius tells of
other olive-trees at Nazareth, and of the cursed fig-
tree, whose stump was remaining above 1500 years.
Not to omit that other fig-tree, (yet standing near
Cairo) which is said to have open'd in two parts, to
receive and protect the Blessed Virgin and Holy Babe,
as she was flying into Egypt ; but is now shew'd
whole again, as Monconys, who saw (but believ'd

nothing of it) tells the story. There is yet there a
tree of the same kind, which measures 17 paces in
circumference : And now in the Aventine Mount
they shew us the *malus Medica*, planted by the hand
of St. Dominic, and another in the monastery at Fundi,
where Thomas Aquinas lived, planted by that saint,
1278. In Congo they speak of trees capable to be
excavated into vessels, that would contain two hundred men a-piece. To which add those superannuated
tilia's now at Basil, and that of Auspurg, under whose
prodigious shade they so often feast, and celebrate
their weddings ; because they are all of them noted
for their reverend antiquity ; that of Basil branching
out 100 paces diameter, from a stem of about 20
foot in circle, under which the German Emperors
have sometimes eaten : And to such trees it seems
they paid divine honours, as the nearest emblems of
eternity, *& tanquam sacras ex vetustate*, as Quintilian
speaks. And like to these might that cypress be,
which is celebrated by Virgil, near to another monument.

5. But we will spare our reader, and refer him
that has a desire to multiply examples of this kind,
to those undoubted records our naturalist mentions in
his 44 chap. lib. 16. where he shall read of Scipio
Africanus's olive-trees ; Diana's lotus ; the ruminal
fig-tree ; under which the bitch-wolf suckl'd the
founder of Rome and his brother; lasting (as Tacitus
calculated) 840 years ; putting out new shoots, presaging the translation of that empire from the Caesarian
line, hapning in Nero's reign. The *ilex*, of prodigious
antiquity, as the Hetruscian inscription remaining on
it imported : But Pausanias in his *Arcadica*, thinks
the Samian *vitex* (of which already) to be one of the

oldest trees growing, and the platan set by Menelaus; to these he adds the Delian palm, co-evous with Apollo himself; and the olive planted by Minerva according to their tradition; the over-grown myrtil; the Vatican and the Holm, and the Tiburtine, and especially that near to Tusculum, whose body was thirty five foot about; besides divers others which he there enumerates in a large chapter: And what shall we conjecture of the age of Xerxes's huge *platanus*, in admiration whereof he staid the march of so many hundred thousand men for so many days; by which the wise Socrates was us'd to swear? And certainly, a goodly tree was a powerful attractive, when that prudent consul, Passienus Crispus, fell in love with a prodigious beech of a wonderful age and stature, which he us'd to sleep under, and would sometimes refresh it with pouring wine at the roots; and that wise Prince Francis the First, as much enamour'd with an huge oak, which he caus'd to be so curiously immur'd at Bourges.

6. We have already made mention of Tiberius's larch, intended to be employ'd about the *Naumachia*, which being one hundred and twenty foot in length, bare two foot diameter all that space, (not counting the top) and was look'd upon as such a wonder, that though it was brought to Rome to be us'd in that vast fabrick, the Emperor would have it kept *propter miraculum*; and so it lay unemploy'd till Nero built his amphitheatre. To this might be added the mast of Demetrius's *Galeasse*, which consisted but of one cedar: And that of the float which wafted Caligulus's obelisks out of Egypt, four fathoms in circumference. We read also of a cedar growing in the island of Cyprus, which was 130 foot long, and 18 in diameter;

and such it seems there are some, yet growing on
Mount Libanus, (tho' so very few in number).   Our
late traveller [1] Mr. Maundrill, affirms himself to have
measur'd one of 12 yards 6 inches in girt, sound, and
no less than thirty yards from the ground, divided
into five limbs, each of which was equal to a great
tree : Of the plane in Athens, whose roots extended
36 cubits farther than the boughs, which were yet
exceedingly large ; and such another was that most
famous tree at Veliternus, whose arms stretch'd out
80 foot from the stem : But these were solid.   Now
if we will calculate from the hollow, besides those
mention'd by Pliny, in the Hercynian forest ; the
Germans had castles in oaks, and (as now the Indians)
had of old some *punti*, or canoos of excavated oak,
which would well contain thirty, some forty persons:
Such were the ancient μονόξυλα, in use yet about
Cephalonia, as Sir George Wheeler observ'd ; and
such the Ἄδρυα Πλοῖα us'd by those of Cyprus : But
what were these to a canoo in Congo, which was
made to hold 200 men ? And the Lician *platanus*
recorded by the naturalist, and remaining long after
his days, had a room in it of eighty one feet in
compass, adorn'd with fountains, stately seats, and
tables of stone ; for it seems it was so glorious a
tree both in body and head, that Licinius Mutianus
(three times consul, and governour of that province)
us'd to feast his whole retinue in it, chusing rather to
lodge in it, than in his golden-roofed palace ; it was
in compass 80 foot, and grew in Asia.   And of later
date, that vast *cerrus* in which an eremit built his
cell and chappel, so celebrated by the noble Fracast-
orius in his poem *Malteide*. cant. 8. stro. 30.

[1] Maundrill's *Journey to Jerusalem*, p. 140.

But for these capacious hollow trees we need go no farther than our own country ; there being (besides that which I mention in Gloucestershire) an oak at Kidlington-green in Oxfordshire, which has been frequently us'd (before the death of the late judge Morton, near whose house it stood) for the immediate imprisonment of vagabonds and malefactors, till they could conveniently be remov'd to the county-gaol : And such another prison Dr. Plot does in his excellent *History of Oxfordshire*, mention out of Ferdinand Hertado in Moravia, to be made out of the trunk of a willow, 27 foot in compass : But not to go out of our promis'd bounds, the learned Doctor speaks of an elm growing on Blechington-Green, which gave reception and harbour to a poor great-belly'd woman, (whom the unhospitable people would not receive into their houses) who was brought to bed in it of a son, now a lusty young fellow.   This puts me in mind of that (I know not what to call it) privilege belonging to a venerable oak, lately growing in Knoll-Wood, near Trely-Castle in Staffordshire, of which (I think) Sir Charles Skrymsher is owner; that upon oath made of a bastard's being begotten within the reach of its boughs shade, (which I assure you at the rising and declining of the Sun, is very ample) the offence was not obnoxious to the censure of either ecclesiastical or civil magistrate.   These, with our historians, I rather mention also for their extravagant use, and to refresh the reader with some variety, than for their extraordinary capacity ; because such instances are innumerable, should we pretend to illustrate this particular with more than needs.

And now I have spoken of elms, and other extra-vagancies of trees ; there stands one (as this curious

EE

observer notes) in Binsey Common, six yards diameter
next the ground, which 'tis conjectur'd has been so
improv'd by raising an earthen bank, or seat about it,
which has caus'd it to put forth into spurs ; it not
being so considerable in the higher trunk.

7. Compare me then with these, that nine fathom'd-
deep tree spoken of by Josephus Acosta; the mastick-
tree seen and measur'd by Sir Francis Drake, which
was four and thirty yards in circuit ; those of
Nicaragua and Gambra, which 17 persons could
hardly embrace : Among these may come in the
cotton-tree describ'd by Dampier.   In India (says
Pliny) *arbores tantae proceritatis traduntur, ut sagittis
superari nequeant*, (and adds, which I think material,
and therefore add also) *haec facit ubertas soli, temperies
caeli, & aquarum abundantia.*   Such were those trees
in Corsica, and near Memphis, &c. recorded by
Theophrastus, &c. and for prodigious height, the two
and three hundred foot unparallel'd palms-royal
describ'd by Captain Ligon, growing in our plantations of the Barbadoes ; or those goodly masts of fir
which I have seen and measur'd, brought from New-
England ; and what Bembus relates of those twenty-
fathom'd Antartic-trees ; or those of which Cardan
writes, call'd Ciba, which rising in their several stems
each of twenty foot in compass, and as far distant
each from other, unite in the bole at fifteen foot
height from the ground, composing three stately
arches, and thence ascending in a shaft of prodigious
bulk and altitude : Such trees of 37 foot diameter (an
incredible thing) Scaliger (his antagonist) speaks of,
ad Gambrae fluvium.   Matthiolus mentions a tree
growing in the Island of Cyprus, which contain'd
130 foot high sound timber : And upon Mount Ætna

in Sicily is a place call'd by them *gli Castayne*, from
three chesnut-trees there standing, where in the cavity
of one yet remaining, a considerable flock of sheep is
commonly folded : Kircher's words are these, as seen
by himself, *et quod forsan* παράδοξον *videri possit, ostendit
mihi viae dux, unius castaneae corticem tantae amplitudi-
nis, ut intra eam integer pecorum grex a pastoribus, tanquam in caula commodissima, noctu includeretur. China
Illust.* p. 185. But this, as I remember, was lately
ruin'd by the direful conflagration about Catanea :
And what may we conceive of those trees in the
Indies, one of whose nuts hardly one man is able to
carry ; and which are so vast, as they depend not like
other fruit, by a stalk from the boughs, but are produc'd out of the very body and stem of the tree, and
are sufficient to feed twenty persons at a meal ? There
were trees found in Brazile, that sixteen men could
hardly fathom about, and the Jesuits caused one of
these to be fell'd, for being superstitiously worship'd
by the savages, which was 120 foot in circumference.
The Mexican Emperor is said to have had a tree in
his garden, under whose shade a thousand men might
sit at a competent distance.

We read of a certain fig in the Charibee Islands,
which emits such large buttresses, that great planks
for tables and flooring are cleft out of them, without
the least prejudice to the tree ; and that one of these
does easily shelter 200 men under them : And in
Nieuhoff's Voyage to the East-Indies, of the *kynti*, a
kind of oak, which yield planks of 4 foot breadth, and
40 in length : Strabo, I remember, *Geog.* l. 15. talks
of fifty horsemen under a tree in India ; his words are
ὥσθ' ὑφ' ἑνὶ δένδρῳ μεσημβρίζειν σκιαζομένους ἱππέας πεντήκοντα,
and of another that shaded five *stadia* at once ; and

in another place of a pine about Ida, which measur'd 24 foot diameter, and of a monstrous height : To these may be added the *arbor de Rays*, a certain tree growing in the East-Indies, which propagates it self into a vast forest (if not hinder'd) by shooting up, and then letting a kind of gummy string to fall and drivle from its branches, which takes root in the ground again, and in this process spread a vast circuit, the single stem of some of which are reported to be no less than fifty foot diameter, a thing almost incredible. To this may be added the *balete* describ'd by Mr. Ray, (*append.* 3d vol.) and what he cites of Melckion Barros, who found trees proof against weapons, resisting the force of any edg'd tool, being of a consisture so hard : But even this, and all we have hitherto produced, is nothing to what I find mention'd in the late *Chinese History* (as 'tis set forth upon occasion of the Dutch Embassy) where they tell us of a certain tree call'd *ciennich* (or the tree of a thousand years) in the province of Suchu, near the city Kien, which is so prodigiously large, as to shrowd 200 sheep under one only branch of it, without being so much as perceiv'd by those who approach it. And to conclude with yet a greater wonder, of another in the province of Chekiang, whose amplitude is so stupendiously vast, as fourscore persons can hardly embrace : These gigantick trees, the Chinese-timber merchants transport on floats, upon which they build huts and little cottages, where they live with their families, floating many thousand miles till all be sold, as Le Compte tells us : In the mean time we must not omit the strange and incredible bulk of some oaks standing lately in Westphalia, whereof one serv'd both for a castle and fort ; and another there which contain'd in height 130 foot, and (as some report, 30 foot diameter ; and another which yielded 100 wane load. I have read of a table of walnut-tree to be seen at St. Nicholas's in Lorrain, which held 25 foot broad, all of a piece, and of competent length and thickness, rarely fleck'd and watered ; Scamozzi the architect reports he saw it : Such a monster that might be, under which the Emperor Fred. the Third held his magnificent feast 1472. For in this recension we will endeavour to give a taste of more fresh observations, and to compare our modern timber with the antient, and that, not only abroad, but without travelling into foreign countries for these wonders.

8. What goodly trees were of old ador'd and consecrated by the Dryads, I leave to conjecture from the stories of our ancient Britains, who had they left records of their prodigies in this kind, would doubtless have furnish'd us with examples as remarkable for the growth and stature of trees, as any which we have deduc'd from the writers of foreign countries ; since the remains of what are yet in being (notwithstanding the havock which has universally been made, and the little care to improve our woods) may stand in fair competition with any thing that antiquity can produce.

9. There is somewhere in Wales an inscription extant, cut into the wood of an old beam, thus,

SEXAGINTA PEDES FUERANT IN STIPITE NOSTRO,
EXCEPTA COMA QUÆ SPECIOSA FUIT.

This must needs have been a noble tree, but not without later parallels ; for to instance in the several species, and speak first of the bulks of some immense

trees ; there was standing an old and decay'd chesnut
at Fraiting in Essex, whose very stump did yield thirty
sizable load of logs ; I could produce you another of
the same kind in Gloucestershire, which contains
within the bowels of it a pretty wainscotted room
inlighten'd with windows, and furnish'd with seats,&c.
to answer the Lician *platanus* lately mention'd.

10. But whilst I am on this period ; see what a *tilia*
that most learn'd and obliging person Sir Tho. Brown of
Norwich describes to me in a letter just now receiv'd.

" An extraordinary large and stately *tilia*, *linden*, or
lime-tree, there groweth at Depeham in Norfolk, ten
miles from Norwich, whose measure is this. The
compass in the least part of the trunk or body about
two yards from the ground, is at least eight yards and
half : about the root nigh the earth, sixteen yards,
about half a yard above that, near twelve yards in
circuit : The height to the uppermost boughs about
thirty yards, which surmounts the famous *tilia* of Zu-
rich in Switzerland; and uncertain it is whether in any
tilicetum, or lime-walk abroad it be considerably exceeded : Yet was the first motive I had to view it not
so much the largeness of the tree, as the general
opinion that no man could ever name it ; but I found
it to be a *tilia femina* ; and (if the distinction of
Bauhinus be admitted from the greater, and lesser leaf)
a *tilia platyphyllos* or *latifolia* ; some leaves being three
inches broad ; but to distinguish it from others in the
country, I call'd it *Tilia colossaea Depehamensis*." Thus
that learned person, from this and the like instance,
(as the reader will find in what follows growing in
our own country ;) I am not apt so much to admire
what is pretended so mightily to exceed the refreshing
shades of some of our oaks, beeches, elms, and other

ample umbrages, if diligently compar'd ; as I am to impute it to what the younger[1] Pliny attributes to mens affecting novelties, that *tanta suarum rerum satietas, aliarumque aviditas.*

A poplar-tree not much inferior to this, he informs me grew lately at Harling by Thetford, at Sir William Gawdy's gate, blown down by that terrible hurrican about four years since.

But here does properly intervene the *linden* of Schalouse in Swisse, under which is a bower compos'd of its branches, capable of containing three hundred persons sitting at ease : It has a fountain set about with many tables, formed only of the boughs, to which they ascend by steps ; all kept so accurately, and so very thick, that the sun never looks into it : But this is nothing to that prodigious *tilia* of Newstadt in the Dutchy of Wirtemberg, so famous for its monstrosity, that even the city it self receives a denomination from it, being called by the Germans NEUSTADT ANDER GROSSEN LINDEN, or Newstadt by the great lime-tree.　The circumference of the trunk is 27 foot 4 fingers : The *ambitus* or extent of the boughs 403 *ferè ;* the diameter from south to north 145, from east to west 119 foot ; set about with divers columns and monuments of stone (82 in number at present, and formerly above an hundred more) which several princes and noble persons have adorn'd, and celebrated with inscriptions, arms and devices, and which, as so many pillars, serve likewise to support the umbragious and venerable boughs : And that even the tree had been much ampler, the ruins and distances of the columns declare, which the rude soldiers have greatly impair'd.

[1] L. 8. ep. 20. ad Gallium.

By the date of the antientest columns yet intire, namely *anno* 1555. may be conjectur'd how goodly a tree it was almost two hundred years since. The inscriptions on the several arms and supporters are as follows.

D. V.H.Z. W. CLL.......... *Graff zu Leuchtenberg.* 1591. 1583. 1575. *Albert von rosenberg Ritter.* 1591. *Wolff Keidel alter Furleutium.* 1555. *Some report he planted it.* *Hans Heinrie vonder Tana.* 1583. *Conrad von Flbeg.* 1575. *Friz Nerter von Hertenek.* 1575. *Wirich von Gemmingen.* 1575. *Bartol......Mot.* 1555. *V. Hans Funk der zeit Burgermeister Die erst.* 1555. *Hans Ulrich Stigelheimer zu Durarhenig Fuctlicher. hr. Hoff-meister*, 1591.

Praesul de Langheim rediens Cisterliae ab urbe
Pyramidem hanc posuit flammis caelestibus auctam.
Sentiat haec etiam numen spirabile toto
Pectore, & illius semper sit munere felix.

Johann. Aht zu Langh. 1601. *Joh. Aht zu Schoenthal.* 1584. *Eberhard von Gimmingen.* 1555. *David von Helmstad Amtman. Graff Fridrich zu Mompelyard. Hans Henrick von Lammestein. Sigismund Signiger. L.H.Z.W.A.* 353. *G.L. Mary Graff au Brandenb.* 1562. *Georg. Ernest Graff zu Henneb. Herr zu Aschaffb.* 1575. *Michel Helmling Stattschreiber.* 1555. *Hans Ulrick von Steine.* 1575. *Daniel von Helmstatt. zu Kappenaw.* 1556....... *Stamel von Reischach* 1575. *Willhelm von Crombach* 1588. *Bernolph von Gammingen.* 1588. *Schweiker Wumbold von Umstatt.* 1591. *Henrich Link Pfarrer zu Uden. Andreas von Oberbach Vorsmeist. zu Neu-statt. Neubrecht Bart Keller zu Leustatt.* 1557....... *Ernberg. Thomas Busch von Schorndorff. Wolffang von Gemmingen* 1588. *Feit Kumeter Forstmeister.* 1551. *and* 1530.

After this we might forbear the naming that at Tillburg near Buda in Hungary, growing in the middle of the street, extending to 62 paces from the

stem, sustain'd by 28 columns : Nor that nearer us,
at Cleves in the Low-Countries, a little without the
entring into the town, cut in 8 faces supported with
pillars, and containing a room in the middle, the
head of the tree curiously shap'd : I say, I need not
have charg'd this paragraph with half these, but to
shew how much more the lime-tree seems to be
dispos'd to be brought into these arborious wonders,
than other trees of slower growth : And yet I am
told of a white-thorn at Worms in Germany, planted
in the centre of the quadrangle of the great church,
whose branches held up with stone, is in circle 50
paces : Several more occur too tedious to recite : But
what is all this, take the most spreading of them, to
what we shall shew, whilst that of Nustradt comes
not yet by forty foot near to the dimensions of an
oak standing lately in Worksop-Park, belonging to
his Grace the Duke of Norfolk, Earl Marshall of
England, spreading almost 3000 yards square, and
under the shade whereof near a thousand horse might
commodiously stand at once.   But, besides this gi-
gantic lime-tree, there is likewise a white-thorn,
brought (as the tradition goes) a small twig, out of
Palestine, *anno* 1470. by Eberhard, first Duke of
Wirtemberk, and planted near Tubing, where he
founded St. Peter's monastery, the branches whereof
being sustain'd by forty columns of stone, is yet a
flourishing tree : 'Tis probable that of Glastenbury is
of this kind, and above a thousand years ancienter, if
the report be true.   At Forti grows a filbert whose
trunk is as big as three mens middles : Near Essling
is a juniper-tree of almost two foot diameter in the
lower trunk, and very tall : These prodigies, with
several more we have from Dr. Faber, physician to

Frederic Duke of Wirtemberg, and collected by the
late industrious Jesuit Schotti in his *Appendix ad
lib. 2. de mirabilibus miscellaneis*.   Nor may here that
goodly birch-tree be forgotten, which growing in one
of the courts of the palace of Augsburgh, is so spreading, as that the branches will cover 365 tables, even
as many as there are days in the year, with its shade,
as Tavernier tells us in his *Travels*.   Mr. Cook, in
his ingenious and useful *Treatise*, mentions a witch-
elm growing within these three or four years in Sir
Walter Baggot's park in the county of Stafford, which
after two men had been five days felling, lay forty
yards in length ; was at the stool seventeen foot
diameter : It broke in the fall fourteen load of wood,
forty eight load in the top : Yielded eight pair of
naves, 8660 foot of boards and planks : It cost ten
pounds seventeen shillings the sawing, the whole
esteem'd 97 tuns : This was certainly a goodly stick.

What other prodigious trees do at present, and of
late abound in that country, may be seen in Dr. Plot's
Natural History ; nay, some planted in the memory
of men of the place, that have grown to a wonderful
procerity : Such was an oak at Narbury, of 15 yards
in girth, which being fell'd, two men at either side
on horse-back could not see one another : And of an
ash of 8 foot diameter, the timber of which was
valued at 30*l.*

11. I am told of a very withy-tree to be seen
somewhere in Barkshire, which is increased to a most
stupendious bulk ; and of two witch-hazel-trees of
prodigious size, growing in Oaksey-Park, belonging
to Sir Edw. Pooles near Malmsbury in Wiltshire ;
not inferior to the largest oaks : But these for arriving
hastily to their *acme* and period, and generally not so

considerable for their use ; I pass to the ash, elm, oak, &c.

There were of the first of these divers which measur'd in length one hundred and thirty two foot, sold lately in Essex : And in the mannor of Horton (to go no farther than the Parish of Ebsham in Surrey, belonging to my brother Richard Evelyn, Esq;) there were elms standing in good numbers, which would bear almost three foot square for more than forty foot in height, which is (in my judgment) a very extraordinary matter. They grow in a moist gravel, and in the hedge-rows.

Not to insist upon beech, which are frequently very large ; there are oaks of forty foot high, and five foot diameter yet flourishing in divers old parks of our nobility and gentry : And firs of 150 foot in height : which is exceeded by one growing in a wood about Bern by almost 100 foot, as Chabrous tells us.

A large and goodly oak there is at Reedham in Sir Richard Berney's park of Norfolk, which I am inform'd was valu'd at forty pounds the timber, and twelve pounds the lopping wood.

12. Nor are we to over-pass those memorable trees which so lately flourished in Dennington Park near Newbury ; amongst which three were most remarkable from the ingenious planter, and dedication (if tradition hold) of the famous English bard, Jeofry Chaucer ; of which one was call'd the King's, another the Queen's, and a third Chaucer's oak. The first of these was fifty foot in height before any bough or knot appear'd, and cut five foot square at the butt-end, all clear timber. The Queen's was fell'd since the wars, and held forty foot excellent timber, straight as an arrow in growth and grain, and cutting

four foot at the stub, and near a yard at the top ;
besides a fork of almost ten foot clear timber above
the shaft, which was crown'd with a shady tuft of
boughs, amongst which, some were on each side
curved like rams-horns, as if they had been so industriously bent by hand. This oak was of a kind so
excellent, cutting a grain clear as any clap-board (as
appear'd in the wainscot which was made thereof)
that a thousand pities it is some seminary of the
acorns had not been propagated, to preserve the
species. Chaucer's oak, though it were not of these
dimensions, yet was it a very goodly tree : And this
account I receiv'd from my most honour'd friend
Phil. Packer, Esq ; whose father (as lately the gentleman his brother) was proprietor of this park : But
that which I would farther remark, upon this occasion,
is, the bulk and stature to which an oak may possibly
arrive within less than three hundred years ; since it
is not so long that our poet flourish'd (being in the
reign of King Edward the Third) if at least he were
indeed the planter of those trees, as 'tis confidently
affirm'd. I will not labour much in this enquiry ;
because an implicit faith is here of great encouragement; and it is not to be conceiv'd what trees of a good
kind, and in apt soil, will perform in a few years ;
and this (I am inform'd) is a sort of gravelly clay,
moisten'd with small and frequent springs. In the
mean while, I have often wish'd, that gentlemen
were more curious of transmitting to posterity, such
records, by noting the years when they begin any
considerable plantation ; that the ages to come may
have both the satisfaction and encouragement by
more accurate and certain calculations. Henry
Ranjovious planted a grove in Ditmarsh, *anno* 1580,

of oak, fir, beech, birch, &c. and erected a stone with
this inscription, (which I mention not for its elegancy,
but example) *an. dom.* 1580, *quercus, abietes, betulas,
&c. plantavit: annum & initium sationis adscribi jussit;
ut earum aetatem exploraret posteritas; quod in omnia
orbis saecula aeternae Divinitati commendat;* as I find it
recorded by that industrious geneologist, Scipio
Amiratus of Florence. But the only instance I know
of the like in our own country, is in the Park at
Althorp in Northamptonshire, the magnificent seat
of the Right Hon. the Earl of Sunderland. I find a
Jewish tradition, cited by the learned Bochart, that
Noah planted the trees (he supposes cedars) of which
he afterwards built the Ark that preserv'd him : Nor
was it esteem'd any diminution for princes themselves
to plant trees with that hand which held the scepter
and reins of empire : So as in the Voorhout of the
Hague, stands a tree plac'd there by the hands of the
Emperor Charles, which is yet in its prime growth,
and no small boast of the good people : But to
proceed.

    13. There was in Cuns-burrow (sometimes belong-
ing to my Lord of Dover) several trees bought by
a cooper, of which he made ten pound per yard for
three or four yards, as I have been credibly assur'd :
But where shall we parallel that mighty tree which
furnish'd the main-mast to the sovereign of our seas,
which being one hundred foot long save one, bare
thirty five inches diameter. Yet was this exceeded
in proportion and use, by that oak which afforded
those prodigious beams that lie thwart her. The
diameter of this tree was four foot nine inches, which
yielded four square beams of four and forty foot long
each of them. The oak grew about Framlingham

in Suffolk ; and indeed it would be thought fabulous but to recount only the extraordinary dimensions of some timber-trees growing in that country ; and of the excessive sizes of these materials, had not mine own hands measur'd a table (more than once) of above five foot in breadth, nine and an half in length, and six inches thick, all intire and clear (not reckoning the slab.)   This plank cut out of a tree fell'd by my grandfather's order, was made a pastry-board, and lay on a frame of solid brick-work at Wotton in Surrey, where it was so placed before the room was finish'd about it, or wall built, and yet abated by one foot shorter, to confine it to the intended dimensions of the place ; for at first, it held this breadth, full ten foot and an half in length : By an inscription cut in one of the sides, it had lain there above an hundred years.   To this may be added, that table of one plank, of above 75 foot long, and a yard broad through the whole length, now to be seen in Dudly-Castle-Hall, which grew in the park, describ'd by Dr. Plot, *Nat. Hist. of Staffordshire.*   Mersennus tells us that the great ship call'd the *Crown,* which the late French King caus'd to be built, has its keel-timber 120 foot long ; and the main-mast 12 foot diameter at the bottom, and 85 in height.

14.   To these I might add a yew-tree in the Church-yard of Crowhurst in the County of Surrey, which I am told is ten yards in compass ; but especially that superannuated yew-tree growing now in Braburne Church-yard, not far from Scots-Hall in Kent ; which being 58 foot 11 inches in the circumference, will bear near twenty foot diameter, as it was measur'd first by my self imperfectly, and then more exactly for me, by order of the late Right

Honourable Sir George Carteret, Vice-Chamberlain
to His Majesty, and late Treasurer of the Navy: Not
to mention the goodly planks, and other considerable
pieces of squar'd and clear timber, which I observ'd
to lie about it, that had been hew'd, and sawn out of
some of the arms only torn from it by impetuous
winds. Such another monster I am inform'd is also
to be seen in Sutton church-yard, near Winchester.
To these we add what we find taken notice of by
the learned, and industriously curious Dr. Plot, in his
Natural History of *Oxfordshire* ; particularly an oak
between Nuneham Courtney and Clifton, spreading
from bough-end to bough-end, 81 foot, shading in
circumference 560 square yards of ground, under
which 2420 men may commodiously stand in shelter.
And a bigger than this near the gate of the water-
walk at Magdalen-Colledge, whose branches shoot
16 yards from the stem ; likewise of another at Ricat
in the Lord Norrey's Park, extending its arms 54
foot, under which 304 horses, or 4374 men may
sufficiently stand : This is that *robur Britannicum* so
much celebrated by the late author of *Dodona's Grove*,
and under which he leans contemplating in the front-
ispiece. But these (with infinite others, which I
am ready to produce) might fairly suffice to vindicate
and assert our proposition, as it relates to modern
examples, and sizes of timber-trees, comparable to any
of the ancients, remaining upon laudable and unsus-
pected records ; were it not great ingratitude to conceal a most industrious, and no less accurate account,
which comes to my hands from Mr. Halton, Auditor
to the Right Honourable the most illustrious and
noble Henry Duke of Norfolk, Earl Marshal of
England.

IN SHEFFIELD LORDSHIP. [1]

15. In the Hall Park, near unto Rivelin, stood an oak which had eighteen yards without bough or knot, and carried a yard and six inches square at the said height, or length, and not much bigger near the root : Sold twelve years ago for 11 *li*. Consider the distance of the place, and country, and what so prodigious a tree would have been worth near London.

In Firth's farm within Sheffield Lordship, about twenty years since, a tree blown down by the wind, made, or would have made two forge-hammer-beams, and in those, and the other wood of that tree, there was of worth, or made 50*li*. and Godfrey Frogat (who is now living) did oft say, he lost 30*li*. by the not buying of it.

A Hammer-beam is not less than 7½ yards long, and 4 foot square at the barrel.

In Sheffield Park, below the mannor, a tree was standing which was sold by one Giffard (servant to the then Countess of Kent) for 2*li*. 10*s*. to one Nich. Hicks; which yielded of sawn wair fourteen hundred, and by estimation, twenty cords of wood.

A wair is two yards long, and one foot broad, six-score to the hundred : So that in the said tree was 10080 foot of boards ; which, if any of the said boards were more than half-inch thick, renders the thing yet more admirable.

In the upper end of Rivelin stood a tree, call'd the Lord's-oak, of twelve yards about, and the top yielded twenty one cord, cut down about thirteen years since.

---

[1] *The Names of the persons who gave intelligence of the particulars are :—*
　　　Edw. Rawson.
　　　Cap. Bullock.
　　　Edw. Morphy, *Wood-ward*.

In Sheffield Park, an. 1646, stood above 100 trees worth 1000*li.* and there are yet two worth above 20 *l.* Still note the place and market.

In the same park, about eight years ago, Ralph Archdall cut a tree that was thirteen foot diameter at the kerf, or cutting place near the root.

In the same park two years since, Mr. Sittwell, with Jo. Magson did chuse a tree, which after it was cut, and laid aside flat upon a level ground, Sam. Staniforth a keeper, and Edw. Morphy, both on horse-back, could not see over the tree one anothers hat-crowns. (And such another was the Marbury oak, mention'd in sect. 10 of this chapter.) This tree was afterwards sold for 20*li.*

In the same park, near the Old Foord, is an oak-tree yet standing, of ten yards circumference.

In the same park, below the Conduit Plain, is an oak-tree which bears a top, whose boughs shoot from the boal some fifteen, and some sixteen yards.

Then admitting $15\frac{1}{2}$ yards for the common, or mean extent of the boughs from the boal, which being doubled, is 31 yards; and if it be imagin'd for a diameter, because the ratio of the diameter to the circumference is $\frac{113}{355}$, it follows $113 : 355 :: 31 : 97 \frac{44}{113}$ yards, which is the circumference belonging to this diameter.

Then farther it is demonstrable in geometry, that half the diameter multiplied into half the circumference produces the area or quantity of the circle, and that will be found to be $754 \frac{347}{452}$ which is 755 square yards *ferè.*

Then lastly, if a horse can be limited to three square yards of ground to stand on (which may seem a competent proportion of three yards long, and one yard broad) then may 251 horses be well said to stand under the shade of this tree. But of the more northern cattle certainly, above twice that number.

Worksopp-Park.

In this park, at the corner of the Bradshaw-rail, lieth the boal of an oak-tree which is twenty nine foot about, and would be found thirty, if it could be justly measur'd; because it lieth upon the ground; and the length of this boal is ten foot, and no arm nor branch upon it.

In the same park, at the white-gate, a tree did stand that was from bough-end to bough-end (that is, from the extream ends of two opposite boughs) 180 foot; which is witness'd by Jo. Magson and Geo. Hall, and measur'd by them both.

Then because 180 foot, or 60 yards is the diameter; 30 yards will be the semidiameter: And by the former analogies

$$113 : 355 :: 60 : 188\tfrac{1}{2}$$
$$\text{and}$$
$$1 : 30 :: 94\tfrac{1}{4} : 2827\tfrac{1}{2}$$

That is, the content of ground upon which this tree perpendicularly drops, is above 2827 square yards, which is above half an acre of ground: And the assigning three square yards (as above) for an horse, there may 942 be well said to stand in this compass.

In the same park (after many hundreds sold, and carried away) there is a tree which did yield quarter-cliff bottoms that were a yard square: and there is of them to be seen at Worksop at this day, and some tables made of the said quarter-cliff likewise.

In the same park, in the place there call'd the Hawks-nest, are trees forty foot long of timber, which will bear two foot square at the top-end or height of forty foot.

If then a square whose side is two foot, be inscribed in a circle, the proportions at that circle are

feet

diameter	2 :	8284
circumference	8 :	8858
area	6 :	2831

And because a tun of timber is said to contain forty solid feet : one of these columns of oak will contain above six[1] tun of timber and a quarter : In this computation taking them to be cylinders, and not tapering like the segments of a cone.

WELBEEK-LANE.

17. The oak which stands in this lane call'd Grindal Oak, hath at these several distances from the ground these circumferences.

	foot	foot inch
at 1	33 :	01
at 2	28 :	05
at 6	25 :	07

The breadth is from bough-end to bough-end (*i.e.*) diametrically 88 foot ; the height from the ground to the top-most bough 81 foot [this dimension taken from the proportion that a *gnomon* bears to the shadow] there are three arms broken off and gone, and eight very large ones yet remaining, which are very fresh and good timber.

88 foot is 29⅓ yards, which being in this case admitted for the diameter of a circle, the square

---

[1] A statutable tun of timber is by some reckon'd 43 foot of solid: and to a load 50.

yards in that circumference will be 676 *ferè*;
and then allowing three yards (as before) for a
beast, leaves 225 beasts, which may possibly
stand under this tree.

But the Lord's Oak, that stood in Rivelin, was in
diameter three yards, and twenty eight inches; and
exceeded this in circumference three feet, at one foot
from the ground.

SHIRE-OAK.

Shire-Oak is a tree standing in the ground late Sir
Tho. Hewet's, about a mile from Worksopp-Park,
which drops into three shires, viz. York, Nottingham
and Derby, and the distance from bough-end to
bough-end is ninety foot and thirty yards.

This circumference will contain near 707 square
yards, sufficient to shade 235 horse.

Thus far the accurate Mr. Halton.

18. Now among such venerable trees (especially
conspicuously plac'd as this last Mr. Halton has nam'd)
should be spared for the most noble and natural
boundaries to great parishes, and gentlemens estates,
famous for which is the chesnut-tree at Tamworth in
Gloucestershire; which has continu'd a signal boundary to that mannor in King Stephen's time, as it stands
upon record: See lib. III. cap. 7, 18. And now
before I shut up these encouraging instances, I am
inform'd by a person of credit, that an oak in Sheffield-
Park, call'd the Ladies Oak, fell'd, contain'd forty two
tun of timber, which had arms that held at least four
foot square for ten yards in length; the body six foot
of clear timber: That in the same park, one might
have chosen above 1000 trees worth above 6000 *li.*
another 1000 worth 4000 *li. & sic de ceteris.* To this

Mr. Halton replies, that it might possibly be meant of the Lords-Oak already mention'd, to have grown in Rivelin : For now Rivelin it self is totally destitute of that issue she once might have gloried in of oaks ; there being only the Hall-Park adjoyning, which keeps up with its number of oaks. And as to the computation of 1000 trees formerly in Sheffield-Park worth 6000 *li.* it is believ'd there were a thousand much above that value ; since in what is now inclos'd, it is evident touching 100 worth a thousand pounds. I am inform'd that an oak (I think in Shropshire) growing lately in a copp'ce of my Lord Cravens, yielded 19 tun and half of timber, 23 cord of fire-wood, 2 load of brush, and 2 load of bark. And my worthy friend Leonard Pinckney, Esq ; lately first clerk of his Majesty's kitchin, did assure me, that one John Garland built a very handsome barn, containing five baies, with pan, posts, beams, spars, &c. of one sole tree, growing in Worksopp-Park. I will close this with an instance which I greatly value, because it is transmitted to me from that honourable and noble person, Sir Edw. Harley : I am (says he) assur'd by an inquisition taken about 300 years since, that a park of mine, and some adjacent woods, had not then a tree capable to bear acorns ; yet, that very park I have seen full of great oaks, and most of them in the extreamest wane of decay. The trunk of one of these oaks afforded so much timber, as upon the place would have yielded 15 *li.* and did compleatly seat with wainscot-pews a whole church : You may please (says he, writing to Sir Rob. Morray) to remember when you were here, you took notice of a large tree, newly fall'n ; when it was wrought up, it proved very hollow and unsound: One of its cavities contain'd two Hogsheads of water : Another was fill'd with better stuff, wax and hony: Notwithstanding all defects, it yielded, besides three tun of timber, 23 cords of wood. But my own trees are but chips in comparison of a tree in the neighbourhood, in which every foot forward, one with another, was half a tun of timber ; it bore 5 foot square, 40 foot long ; it contain'd 20 tun of timber, most of it sold for 20s. per tun ; besides that, the boughs afforded 25 cords of fuel-wood : This was call'd the Lady-Oak : Is't not pity such goodly creatures should be devoted to Vulcan ? &c. So far this noble gent. to which I would add *dirae*, a deep execration of iron-mills, and I had almost said iron-masters too,

Quos ego ; sed motos praestat componere.............

for I should never finish, to pursue these instances through our once goodly magazines of timber for all uses, growing in this our native country, comparable (as I said) to any we can produce of elder times; and that not only (though chiefly) for the encouragement of planters, and preservers of one of the most excellent and necessary materials in the world for the benefit of man ; but to evince the continu'd vigor of nature, and to reproach the want of industry in this age of ours ; and (that we may return to the argument of this large chapter) to assert the procerity and stature of trees from their very great antiquity : For certainly, if that be true, which is by divers affirmed concerning the *quercetum* of Mamre (where the Patriarch en- tertain'd his angelical guests) recorded by Eusebius to have continued till the time of Constantine the Great, we are not too prejudicately to censure what has been produc'd for the proofs of their antiquity ; nor for

my part do I much question the authorities : But let
this suffice ; what has been produced being not only
an historical speculation of encouragement and use,
but such as was pertinent to the subject under consideration, as well as what I am about to add concerning the texture, and similar parts of the body of trees,
which may also hold in shrubs, and other lignous
plants ; because it is both a curious, and rational
account of their anatomization, and worthy of the
sagacious enquiry of that learned person, the late Dr.
Goddard, as I find it entered amongst other of those
precious collections of this illustrious Society.

19. The trunk or bough of a tree being cut transversely plain and smooth, sheweth several circles or
rings more or less orbicular, according to the external
figure, in some parallel proportion, one without the
other, from the centre of the wood to the inside of
the bark, dividing the whole into so many circular
spaces.   These rings are more large, gross, and distinct in colour and substance in some kind of trees,
generally in such as grow to a great bulk in a short
time, as fir, ash, &c. smaller or less distinct in those
that either not at all, or in a longer time grow great;
as quince, holly, box, *lignum-vitae*, ebony, and the
like sad colour'd and hard woods ; so that by the
largeness or smallness of the rings, the quickness or
slowness of the growth of any tree may perhaps at
certainty be estimated.

These spaces are manifestly broader on the one
side, than on the other, especially the more outer, to
a double proportion, or more ; the inner being near
an equality.

It is asserted, that the larger parts of these rings are
on the south and sunny side of the tree (which is very

rational and probable) insomuch, that by cutting a tree transverse, and drawing a diameter through the broadest and narrowest parts of the rings, a meridian line may be described.

The outer spaces are generally narrower than the inner, not only in their narrower sides, but also on their broader, compared with the same sides of the inner: Notwithstanding which, they are for the most part, if not altogether, bigger upon the whole account.

Of these spaces, the outer extremities in fir, and the like woods, that have them larger and grosser, are more dense, hard, and compact ; the inner more soft and spungy ; by which difference of substance it is, that the rings themselves come to be distinguished.

According as the bodies and boughs of trees, or several parts of the same, are bigger or lesser, so is the number, as well as the breadth of the circular spaces greater or less ; and the like, according to the age, especially the number.

It is commonly, and very probably asserted, that a tree gains a new one every year. In the body of a great oak in the New-Forest, cut transversely even (where many of the trees are accounted to be some hundreds of years old) three and four hundred have been distinguished. In a fir-tree, which is said to have just so many rows of boughs about it, as it is of years growth, there has been observed just one less, immediately above one row, than immediately below. Hence some probable account may be given of the difference between the outer, and the inner parts of the rings, that the outermost being newly produced in the Summer, the exterior superficies is condens'd in the Winter.

20. In the young branches and twigs of trees there is a pith in the middle, which in some, as ash, and especially elder, equals, or exceeds in dimensions the rest of the substance, but waxes less as they grow bigger, and in the great boughs and trunk scarce is to be found : This gives way for the growth of the inward rings, which at first were less than the outer (as may be seen in any shoot of the first year) and after grow thicker, being it self absum'd, or perhaps converted into wood ; as it is certain cartilages or gristles are into bones (in the bodies of animals) from which to sense they differ even as much as pith from wood.

These rings or spaces appearing upon transverse section (as they appear elliptical upon oblique and straight lines upon direct section) are no other than the extremities of so many integuments, investing the whole tree, and (perhaps) all the boughs that are of the same age with any of them, or older.

The growth of trees augmentation in all dimensions is acquired, not only by accession of a new integument yearly, but also by the reception of nourishment into the pores and substance of the rest, upon which they also become thicker ; not only those towards the middle, but also the rest, in a thriving tree : Yet the principal growth is between the bark and body, by accession of a new integument yearly, as hath been mentioned : Whence the cutting of the bark of any tree or bough round about, will certainly kill it.

The bark of a tree is distinguished into rings, or integuments, no less than the wood, though much smaller or thinner, and therefore not distinguishable, except in the thick barks of great old trees, and toward the inside next the wood ; the outer parts

drying and breaking with innumerable fissures, growing wider and deeper, as the body of the tree grows
bigger, and mouldering away on the outside.

Though it cannot appear by reason of the continual
decay of it, upon the account aforesaid ; yet it is
probable, the bark of a tree hath had successively as
many integuments as the wood ; and that it doth
grow by acquisition of a new one yearly on the
inside, as the wood doth on the outside ; so that the
chief way, and conveyance of nourishment to both
the wood and the bark, is between them both.

The least bud appearing on the body of a tree,
doth as it were make perforation through the several
integuments to the middle, or very near ; which part
is as it were, a root of the bough into the body of
the tree ; and after becomes a knot, more hard than
the other wood : And when it is larger, manifestly
shewing it self also to consist of several integuments,
by the circles appearing in it, as in the body : More
hard, probably, because straitned in room for growth ;
as appears by its distending, buckling as it were, the
integuments of the wood about it ; so implicating
them the more ; whence a knotty piece of wood is so
much harder to cleave.

It is probable, that a cyon or bud, upon graffing,
or inoculating, doth as it were, root it self into the
stock in the same manner as the branches, by producing a kind of knot.   Thus far the accurate Doctor.

21. To which permit me to add only (in reference
to the circles we have been speaking of) what another
curious enquirer suggests to us ; namely, that they
are caus'd by the pores of the wood, through which
the sap ascends in the same manner as between the
wood and the bark ; and that in some trees, the bark

adheres to the wood, as the integuments of wood cleave to one another, and may be separated from each other as the bark from the outwardmost ; and being thus parted, will be found on their outsides to represent the colour of the outermost, contiguous to the bark ; and on the inner sides, to hold the colour of the inner side of the bark, and all to have a deeper or lighter hue on their inner side, as the bark is on that part more or less tinged ; which tincture is suppos'd to proceed from the ascendent sap.   Moreover, by cutting the branch, the ascending sap may be examin'd as well as the circles : It is probable, the more frequent the circles, the larger, and more copiously the liquor will ascend into it ; the fewer, the sooner descend from it.   That a branch of three circles cut off at Spring, the sap ascending, will be found at Michaelmass ensuing ; cut again in the same branch, or another of equal bigness, to have one more than it had at Spring ; and either at Spring or Fall to carry a circle of pricks next the bark, at other seasons a circle of wood only next it.   But here the comparison must be made with distinction ; for some trees do probably shoot new tops yearly till a certain period, and not after ; and some have perhaps their circles in their branches decreased from their bodies to the extremity of the branch, in such oeconomy and order ; that (for instance) an apple-tree shoot of this year has one circle of pricks or wood less, than the graft of two years growth ; and that of two years growth, may the next year have one circle more than it had the last year ; but this only till that branch shoot no more grafts, and then 'tis doubtful whether the outmost twig obtain any more circles, or remain at a stay, only nourished, not augmented in the circles. It would also be enquir'd, whether the circles of pricks increase not till Midsummer and after, and the circles of wood from thence, to the following Spring? But this may suffice, unless I should subjoin

22. The vegetative motion of plants, with the diagrams of the Jesuit Kircher, where he discourses of their stupendious magnetisms, &c. could there any thing material be added to what has already been so ingeniously enquired into by the learned Dr. Grew in his *Anatomy of Vegetables*, and that of *Trunks ;* where experimentally, and with extraordinary sagacity, he discusses the present subject (with entire satisfaction of the inquisitive reader) beginning at the seeds, to the formation of the root, trunk, branches, leaves, flower, fruit, &c. where you have the most accurate descriptions of the several vessels, for sap, air, juices, with the stupendious contexture of all the organical parts ; and than which there can be nothing more fully entertaining : So that what Dr. Goddard, and other ingenious men have but conjecturally hinted, is by this inquisitive person (and that of the excellent Malpighius) evinced by autoptical experience, and profound research into their anatomy. To all which we may by no means forget the most Lincean inspector Mr. Ant. Van Leeuwenhock, concerning the barks of trees, which he affirms, and experiment-ally convinces, that that integument, namely, the bark, was produced from the wood, and not the wood from the bark. But this discourse, together with the microscopical figure, (being too long to be here inserted) refers to that most industrious person's letter, *Transact*. Numb. 296. p. 1843. Let us there-fore proceed to the felling.

23. It should be in this *status*, vigour and perfection of trees, (which for the oak I take to be about the age of 50, or 'twixt that and 60 years growth, where the soil is natural) that a felling should be celebrated; since whilst our woods are growing it is pity, and indeed too soon; and when they are decaying, too late. I do not pretend that a man (who has occasion for timber) is obliged to attend so many ages e're he fell his trees; but I do by this infer, how highly necessary it were, that men should perpetually be planting; that so posterity might have trees fit for their service of competent, that is, of a middle growth and age, which it is impossible they should have, if we thus continue to destroy our woods, without this providential planting in their stead, and felling what we do cut down, with great discretion, and regard of the future.

I know it is an objection, or rather an unreasonable excuse of the slothful neglect of successive and continual planting, upon so tedious an expectation of what is not likely to be timber in our time: But as this is quite otherwise, (provided men would be early at the work) they might have sufficient of their own planting, (nay, from the very rudiment and seeds) abundantly to recompence their patience and attendance, living to the age men usually attain, by the common course of nature; with how much more improvement to their children and posterity? and this minds me of what's reported of the Emperor Maximilian the IId. That by chance finding an ancient husbandman setting date-stones, asks him what his meaning was to plant a tree that required an hundred years before it bare any fruit? Sir, replies the good man, I have children, and they may have

more come after them. At which the Emperor was
so well pleas'd, that he gave him an hundred florins.
Was not this like that of Laertes to Ulysses?

But before we go farther with the history of the
stature and magnitude of trees, we are not to conclude as if all those trees and plants, which arrive to
that enormous stature and bulk we have mention'd,
were not to be found in other countries, both of the
same, and other species; but that even of those
exoticks, and divers of our own, which seem pigmies
and dwarfs, compared to those giants in their native
climate, are so much greater than in ours; since we
find what we account but shrubs, are divers of them
well-grown trees, and prosper into useful timber;
such as juniper, (emulating the tall cedar) sabine,
tamarisk, cornel, *philyrea*, granade, *lentiscus*, *thuya*,
laurel, bays, and even rosemary, (and other *frutexes*
and lignous plants) superior in growth and stature,
(than with us) where they spontaneously emerge.
Thus not only the white-mulberry wonderfully out-
strips ours, but those of much smaller stature; as the
arbutus, growing on Mount Athos, which became a
spreading tree; so the cypress in Candy to timber,
fit for vast beams, and planks of 4 foot breadth: The
larch overtopping the fir; nay, the myrtil with us
but a bush, make staves for spears; the oleander, &
humilis genista; nay, the rhododendron posts and
rafters; and even herbaceous suffrutages, and amongst
the culinary furniture; a grain of mustard springing
to a tree, whose branches afford harbour to the birds
of the air; and the very hyssop, for a stalk that carried a sponge to the mouth of our Blessed Lord on
the Cross. We are told by Josephus, in Macherontis's
reign, there was a plant of rue growing, and was equal

for height and thickness, to any fig-tree, as was still remaining to the time of Herod, and would have stood longer, had not the Jews cut it down, Jos. *Antiq. Bell. Jud.* lib. VII. cap. 38. How these, and indeed all other vegetables differ in the north, from those of the south, growing on the same mountain, Monsieur Brenier has shewn us ; some nipt and starv'd with that *penetrabile frigus* and scorching heat, quite changing almost their very nature and constitution ; some of them dry, and yielding nothing but leaves, others of the same species are gummy, juicy, and succulent : The *lentiscus* yields mastich in Cio ; in Italy, the oak bears galls ; and the *fraxinus* exsudes manna in Calabria : Thus do *caelum* and *solum* govern the vegetable kingdom, for the mutual supply of the most useful productions, especially that of the forest ; without which, there could be no commerce in the world ; for so has Providence ordain'd. Let us now proceed with felling.

24. Such as we shall perceive to decay, should first be pick'd out for the ax : and then those which are in their state, or approaching to it ; but the very thriving, and manifestly improving, indulg'd as much as possible. But to explore the goodness and sincerity of a standing tree, is not the easiest thing in the world : We shall anon have occasion to mention my Ld. Bacon's experiment to detect the hollowness of timber : But there is doubtless none more infallible, than the boring it with a midling piercer made auger-fashion, and by frequent pulling out, and examining what substance comes along with it, as those who bore the earth to explore what minerals the place is impregnated with, and as sound cheeses are tasted : Some again there are who by digging a little about

the roots, will pronounce shrewdly concerning the
state of a tree ; and if they find him perish'd at the
top (for trees die upward, as men do from the feet) be
sure the cause lies deep, for 'tis ever a mark of great
decay in the roots.   There is also a swelling vein,
which discovers it self eminently above the rest of
the stem, though like the rest invested with bark,
and which frequently circles about and embraces the
tree, like a branch of ivy, which is an infallible indication of hollowness and hypocrisy within.

25. The time of the year for this destructive work
is not usually till about the end of April (at which
season the bark does commonly rise freely) though
the opinions and practice of men have been very
different : Vitruvius is for an autumnal fall ; others
advise December and [1] January : Cato was of opinion
trees should have first born their fruit, or least, not
till full ripe, which agrees with that of the architect;
who begins his fell from the commencement of
Autumm to the Spring, when Favonius begins to
spire ; and his reason is, for that from thence, during
all the Summer, trees are as it were going with child,
and diverting all their nourishment to the embryo,
leaves, and fruit, which renders them weak and
infirm : This he illustrates from teeming women,
who during their pregnancy are never so healthful,
as after they are delivered of their burden, and abroad
again : And for this reason (says he) those merchants,
who expose slaves to sale, will never warrant one that
is with child : The buyer was (it seems) to stand to
the hazard.   Thus he : But I remember *Monsieur*
Perrault in his pompous edition of our author, and

[1] Post ortum Pleiadum a die 6 kal. Jan. usque ad Arcturi ortum, scil. 8 kal. Octob.
Veget. rei milit. l. 5. c. 9.

learned notes upon this chapter, reproves the instance, and corrects the text, *a disparatione procreationis, &c.* to *ad disparationem, &c.* affirming that women are never more sound and healthy than when they are pregnant ; the nutrition deriv'd to the infant, being (according to him) no diminution or prejudice to the mother ; as being but the consumption of that humidity, which enfeebles the bearing woman, and thence infers, that the comparison cannot hold in trees, which become so much stronger by it : But to insist no longer on this ; there is no doubt, that whilst trees abound in over-much, crude, and super-fluous moisture (though it may, and do contribute to their production and fertility, for which reason Lucina was invok'd by parturient women) they are not so fit for the ax as when being discharg'd of it, and that it rises not in that quantity as to keep on the leaves and fruit, those laxed parts and vessels by which the humour did ascend, grow dry and close, and are not so obnoxious to putrefaction, and the worm : Hence it is that he cautions us to take notice of the moon's decline, because of her dominion over liquids, and directs our woodman (some days before he fells downright) to make the gash or overture, *usque ad mediam medullam*, to the end the whole moisture may exstil ; for that not only by the bark (which those who resemble trees to animals will have to be analogous to arteries) does the juice drain out ; but by that more fatty and whiter substance of the wood it self, immediately under the bark (and which our carpenters call the sap, and therefore hew away, as subject to rot) which they will have to be the veins : It is (say they) the office of these arteries of bark, receiving nourishment from the roots, to derive

it to every part of the tree, and to remand what is
crude and superfluous by the veins to the roots again;
whence, after it has been better digested, it is made
to ascend a second time by the other vessels in perpetual circulation ; and therefore necessary so deep
an incision should be made as may serve to exhaust
both the venal and arterial moisture : But for this
nice speculation, I refer the curious to the already
mention'd Dr. Grew, and to the learned Malpighius,
who have made other, and far more accurate observations upon this subject : In the mean time, as to
that of the worm in timber-trees, and their rotting,
sometimes within, and sometimes without ; observe
that such as gape and rift outwardly, (as does that of
the oak, when fell'd) the sap thereby let out, the
timber and heart within is found to be much more
solid than that of the chesnut and other trees who
keep the moisture within (however seeming sound
outwardly) the timber is frequently extremely rotted
and perish'd : Lastly, concerning the bark, though
some are for stripping it, and so to let the tree stand
till about Mid-June, to preserve it from the worm
(all which time it will put forth leaves, and seemingly
flourish) yet that which is unbark'd, is obnoxious to
them, contracts somewhat a darker hue, which is the
reason so many have commended the season when it
will most freely strip) yet were this to be rather
consider'd for such trees as one would leave round,
and unsquar'd; since we find the wild oak, and many
other sorts, fell'd over-late, and when the saps begins
to grow proud, to be very subject to the worm ;
whereas, being cut about Mid-Winter, it neither
casts, rifts, nor twines ; because the cold of the
Winter does both dry, and consolidate ; whiles in

Spring, and when pregnant, so much of the virtue goes into the leaves and branches : Happy therefore were it for our timber, some real invention of tanning without so much bark (as the Honourable Mr. Charles Howard has most ingeniously offer'd) were become universal, that trees being more early felled, the timber might be better season'd and condition'd for its various uses.   But as the custom is, men have now time to fell their woods, even from Mid-winter to the Spring ; but never any after the Summer Solstice: And now we speak of tanning, they have in Jamaica the mangrave, olive, and a third whose barks tan much better than do ours in England ; so as in six weeks the leather is fit to be employ'd to any use: They have likewise there a tree, whose berries wash better and whiter than any Castile-soap.

26. Then for the age of the moon, it has religiously been observ'd ; and that Diana's presidency *in silvis* was not so much celebrated to credit the fictions of the poets, as for the dominion of that moist planet, and her influence over timber : However experienc'd men commend the felling soon after a full-moon, and so during all the decrease, and so to let the tree lie at least 3 months, to render the timber strong and [1] solid: For my part, I am not so much inclin'd to these criticisms, that I should altogether govern a felling at the pleasure of this mutable lady ; however there is doubtless some regard to be had,

[2] Nor is't in vain signs fall and rise to note.

Whilst as to other more recondit and deep astrological observations, minute and scrupulous, perhaps not

See Macrob. *Sat.* lib. VII. cap. 6.
[2] Nec frustra signorum obitus speculamur, & ortus.

altogether to be rejected, both as to the various configurations of the superior bodies, and operation on
both vegetable and sensitive, especially as to the
growth of fruit, sowing, planting and cultivating :
(Indicating the proper seasons, according to the
access and recess of the greater luminaries, through
the zodiaque): It were ingratitude to impute it all to
the superstition of the Ancients, or the total ignorance of causes in those great and learned men (such
as Hesiod, Virgil, Cato, Varro, Columella, Pliny,
and the rest) who have so freely left us these lessons ;
doubtless from their long experience, and extraordinary penetration and enquiry into nature : Let the
curious then (for his better satisfaction) consult that
learned treatise of *Judicial Astrology*, written by Sir
Christopher Heydon.

In the mean time the old rules are these :

Fell in the decrease, or four days after conjunction
of the two great luminaries ; some the last quarter of
it ; or (as Pliny) in the very article of the change, if
possible ; which hapning (saith he) in the last day of
the Winter Solstice, that timber will prove immortal :
At least should it be from the twentieth to the thirtieth
day, according to Columella : Cato four days after the
full, as far better for the growth, nay oak in the
Summer : But all vimineous trees *silente lunâ* ; such
as sallows, birch, poplar, &c. Vegetius for ship-timber,
from the fifteenth to the twenty-fifth ; the moon as
before ; but never during the increase, trees being
then most abounding with moisture, which is the
only source of putrefaction : And yet 'tis affirm'd
upon unquestionable experience, that timber cut at
any season of the year, in the old moon, or last
quarter, when the wind blows westerly, proves as

sound and good as at any other period whatsoever ; nay, all the whole Summer long, as in any month of the year ; (especially trees that bear no fruit.) Theophrastus will have the fir, pine and pitch-tree fell'd when first they begin to bud : I enumerate them all, because it may be of great use on some publick emergencies.

27. Then for the temper, and time of day : The wind low, neither east nor west (but west of the two) the east being most pernicious, and exposing it to the worms ; and for which the best cure is, the plentiful sobbing it in water ; neither in frosty, wet, or dewy weather ; and therefore never in a fore-noon, but when the season has been a good while dry and calm; for as the rain sobs it too much, so the wind closes and obstructs the moisture from ousing out. Lastly, touching the species : Fell fir when it begins to spring ; not only because it will then best quit its coat and strip ; but for that they hold it will never decay in water ; which howsoever Theophrastus deduces from the old bridge made of this material over a certain river in Arcadia, cut in this season, is hardly sufficient to satisfie our enquiry.

28. Previous to this work of felling is the advice of our countryman Markham, and it is not to be rejected : Survey (saith he) your woods as they stand, immediately after Christmas, and then divide the species in your mind ; (I add rather in some note-book, or tablets) and consider for what purposes every several kind is most useful, which you may find in the several chapters of this discourse under every head. After this reckon the bad and good together, so as one may put off the other, without being forc'd to glean your woods of all your best timber. This done

(or before) you shall acquaint your self with the
marketable prices of the countrey where your fell is
made, and that of the several sorts ; as what so many
inches or foot square, and long, is worth for the
several employments : What planks, what other scantlings, for so many spoaks, naves, rings, pales, poles,
spars, &c. as suppose it were ash, to set apart the
largest for the wheel-wright, the smallest for the
cooper, and that of ordinary scantling for the ploughs,
and the brush to be kidded and sold by the hundred,
or thousand, and so all other sorts of timber, viz.
large, middling stuff, and poles, &c. allowing the
waste for the charges of felling, &c. all which you
shall compute with greater certainty, if you have
leisure, and will take the pains to examine some of
the trees either by your own fathom ; or (more
accurately) by girting it about with a string, and so
reducing it to the square, &c. by which means you
may give a near guess : Or, you may mark such as
you intend to fell ; and then begin your sale about
Candlemas till the Spring ; before which you must
not (according as our custom is) lay the ax to the
root ; though some for particular employments, as for
timber to make ploughs, carts, axle-trees, naves, har-
rows, and the like husbandry-tools, do frequently cut
in October.

Being now entering with your workmen, one of
the first, and most principal things, is, the skilful
disbranching of the boal of all such arms and limbs
as may endanger it in the fall, wherein much forecast
and skill is required of the woodman ; so many excellent trees being utterly spoiled for want of this
only consideration : And therefore in arms of timber,
which are very great, chop a nick under it close to

the boal, so meeting it with the downright strokes, it will be sever'd without splicing.

29. We have shewed why some, four or five days before felling, bore the tree cross-way, others cut a kerf round the body, almost to the very pith, or heart, and so let it remain a while ; by this means to drain away the moisture, which will distill out of the wounded veins, and is chiefly proper for the moister sort of trees : And in this work the very ax will tell you the difference of the sex ; the male being so much harder and browner than the female : But here (and whereever we speak thus of plants) you are to understand the analogical, not proper distinctions.

30. But that none may wonder why in many authors of good note, we find the fruit-bearers of some trees call'd males, and not rather females, as particularly the cypress, &c., this preposterous denomination had (I read) its source from very ancient custom, and was first begun in Ægypt (Diodorus says in Greece) where we are told, that the father only was esteem'd the sole author of generation ; the mother contributing only receptacle, and nutrition to the off-spring, which legitimated their mixtures as well with their slaves as free-women : And upon this account it was, that even trees bearing fruit, were amongst them reputed males, and the sterile and barren ones for females ; and we are not ignorant how learnedly this doctrine has been lately reviv'd by some of our most celebrated physicians : But since the same arguments do not altogether quadrate in trees, where the coition is not so sensible (whatever they pretend of the palms, &c. and other amorous intertwining of roots) in my opinion we might with more reason call that the female which bears any eminent fruit, seed

or egg (from whence animals, as well as trees, not
excepting man himself, as the learned Steno, Swamer-
dam and others have, I think, undeniably made it
out) and them males who produce none : But sometimes too the rudeness, or less asperity of the leaves,
bark and grain, nay their medical operations, may
deserve the distinction; to which Aristotle adds branch-
iness, less moisture, quick maturity, &c. l. 1. *de Pl.*
c. 3. All which seems to be most conspicuous in
plum-trees, hollies, ashes, quince, pears, and many
other sorts ; not to insist on such as may be compelled
even to change, as it were, their sex, by graffing and
artificial improvements : For whatever we are told of
such evident distinction of sexes in some, ([1] *mala
Medica*, &c.) I look upon it as hapning rather through
some accidental protrusion, artificial exuberance or
depression, than constant and natural : But I return to

31. Felling, which should be to leave the stools as
close to the ground as possible may be, especially if
you pesign a renascency from the roots ; unless you
will grub for a total destruction, or the use of that
part we have already mention'd, so far superior in
goodness to what is more remote from the root, and
besides the longer you cut and convert the timber,
the better for many uses. Some are of opinion, that
the seedling oak should never be cut to improve his
boal ; because, say they, it produces a reddish wood
not acceptable to the workman ; and that the tree
which grows on the head of his mother does seldom
prove good timber ; It is observ'd indeed, that one
foot of timber near the root (though divers I know

---

[1] Maranch, l. 11. c. 11. maris enim pomum ad natum habet quoddam veluti
infantis genitale ejusdem cum pomo corticis & coloris ; feminae muliebre pu-
dendum ad veram ejus effigiem efformatum videtur, quo simile magis sculptor
non fingat.

who otherwise opine) and (which is the proper *kerfe*, or cutting place) is worth three farther off : And haply, the successor is more apt to be tender, than what was cut off to give it place ; but let this be enquir'd into at leisure : If it be a Winter-fell, for fuel, prostrate no more in a day than the cattle will eat in two days, I mean of the browsewood, and when that's done, kid, and set it up an end, to preserve it from rotting.

32. Dr. Plot recommends the disbranching to be done in the Spring before felling, whilst the tree is standing, that is, from May to Michaelmas, and so to let it continue till the next Spring, and disburthen them when fell'd, as the custom is in Staffordshire, and the North ; for exceedingly contributing to a dry seasoning, freeing it from the attack of worms and other accidental corruption; and thinks that the prejudice accruing thereby, as to the tanner, (in regard of the more difficult excortication,) is no way to be put in balance with the advantage and improvement of the timber for paling, building of ships and houses, &c.   Accounting this method of that universal importance, as to merit the deliberation of a parliament : In the mean while, by whatever method you proceed as to this ; when once a tree is prostrate, and the bark stripp'd off, let it so be set, as it may be best dry; then cleanse the boal of the branches which were left, and saw it into lengths for the squaring, to which belong the measure, and girth (as our workmen call it) which I refer to the buyer, and to many subsidiary books lately printed, wherein it is taught by a very familiar calcule mechanical and easy method.

33. But by none, in my apprehension, set forth, in

JJ

a more facile and accurate way than what that industrious mathematician Mr. Leybourn has publish'd, in
his late *Line of proportion made easy*, and other his
labours ; where he treats as well of the square as the
round, as 'tis applicable to boards and superficials, and
to timber which is hew'd, or less rough, in so easie
a method, as nothing can be more desired.   I know
our ordinary carpenters, &c. have generally upon
their rulers a line, which they usually call Gunter's
line ; but few of them understand how to work from
it as they should : And divers country gentlemen,
stewards and woodmen, when they are to measure
rough timber upon the ground, confide much to the
girt, which they do with a string at about four, or
five foot distance from the root or great extream : Of
the strings length, they take a quarter for the true
square, which is so manifestly erroneous, that thereby they make every tree so measur'd, more than a fifth
part less than really it is.   This mistake would therefore be reform'd ; and it were (I conceive) worth the
seller's while, to inspect it accordingly : Their argument is, that when the bark of a tree is stripped, and
the body hew'd to a square, it will then hold out
no more measure ; that which is cut off being only
fit for fuel, and the expence of squaring cost more
than the chips are worth.   To convince them of
this error, I shall refer and recommend them to the
above-nam'd author: And to what the industrious
Mr. Cooke has so mathematically demonstrated :
Where also of taking the altitude of trees the better
to judge of the worth of them, with the measuring
of wood-lands, &c. together with necessary calculations for the levelling of ground, and removing of
earth, drawing of plots and figures ; all which are

very conducible to the several arguments of this silvan
work. But to proceed.

34. If you are to remove your timber, let the dew
be first off, and the south-wind blow before you draw
it : Neither should you by any means put it to use
for three or four months after, (some not as many
years) unless great necessity urge you, as it did
Duilius, who in the Punic War, built his fleet of
timber before it was season'd, being not above two
months from the very felling to the launching : And
as were also those navies of Hiero after forty days ;
and that of Scipio, in the third Carthaginian War,
from the very forest to the sea. July is a good time
for bringing home your fell'd timber ; But concerning the time and season of felling, a just treatise
might be written : Let the learned therefore consult
Vitruvius particularly on this subject, l. 2. c. 19.
Also M. Cato, c. 17. Plin. l. 16. c. 31. Constantinus
and Heron. l. 3. *de rr. veget.* l. 4. c. 35. Columella
l. 3. c. 2, but especially the most ample Theophrastus
φυτῶν ἱστορίας, l. 5. Note, that a tun of timber is forty
solid feet, a load, fifty.

35. To make excellent boards and planks, 'tis the
advice of some, you should bark your trees in a fit
season, and so let them stand naked a full year before
the felling ; and in some cases, and grounds, it may
be profitable : But let these, with what has been
already said in the foregoing chapters of the several
kinds, suffice for this article : I shall add one adver-
tisement of caution to those noble persons, and others
who have groves and trees of ornament near their
houses, and in their gardens in London, and the
circle of it; especially, if they be of great stature,
and well grown; such as were lately the groves in

the several Inns of Court; nay, even that (compara-
tively, new plantation) in my Lord of Bedford's [1]
garden, &c. and where-ever they stand in the more
interior parts of this city; that they be not over-
hasty, or by any means persuaded to cut down any
of their old trees, upon hope of new more flourishing
plantations; thickning, or repairing deformities; because they grew so well when first they were set : it
is to be consider'd how exceedingly that pernicious
smoak of the sea-coal is increas'd in, and about London since they where first planted, and the buildings
invironing them, and inclosing it in amongst them,
which does so universally contaminate the air, that
what plantations of trees shall be now begun in any
of those places, will have much ado, great difficulty,
and require a long time to be brought to any tolerable
perfection : Therefore let them make much of what
they have; and tho' I discourage none, yet I can
animate none to cut down the old.

36. And here might now come in a pretty speculation, what should be the reason after general fellings,
and extirpations of vast woods of one species, the next
spontaneous succession should be of quite a different
sort ? We see indeed something of this in our gardens
and corn-fields (as the best of poets witnesses,) but that
may be much imputed to the alteration, by improvement, or detriment of the soil and other accidents :
Whatever the cause may be, since it appears not from
any universal decay of nature (sufficiently exploded) I
shall only here produce matter of fact, and that it
ordinarily happens. As in some goodly woods formerly belonging to my grandfather that were all of oak ;

---

[1] Since the first publication of this discourse, most of those groves and trees
have been cut down, to give place for buildings, and turn'd into streets.

after felling, they universally sprung up beech; and 'tis
affirmed, by general experience, that after beech, birch
succeeds ; as in that famous wood at Darnway on the
river Tindarne, in the province of Moray in Scotland,
where nothing had grown but oak in a wood three
miles in length, and happily more southerly, it might
have been beech, and not birch till the third degrad-
ation.   Birches familiarly grow out of old and decay'd
oaks ; but whence this sympathy and affection should
proceed, is more difficult to resolve, in as much as we
do not detect any so prolifical and eminent seed in that
tree.   Some accidents of this nature may be imputed
to the winds, and the birds who frequently have been
known to waft, and convey seeds to places widely
distant, as we have touch'd in the chapter of firs, &c.
sect. 4.   Holly has been seen to grow out of ash, as
ash out of several trees, especially haw-thorn ; nay, in
an old rotten ash-stump, in a place where no ashes at
all grew by many miles in the whole county : And I
have had it confidently asserted by persons of undoubted truth, that they have seen a tree cut in the middle,
whose heart was ash-wood, and the exterior part oak,
and this in Northamptonshire : And why not as well
(though with something more difficulty) as through a
willow, whose body (as is noted) it has been observed
to penetrate even to the earth ? detruding the willow
quite out of its place, of which a pretty emblem might
be conceiv'd : But I pursue these instances no farther,
concluding this chapter with the Norway engine, or
saw-mill, to be either moved with the force of water,
or wind, &c. for the more expedite cutting, and converting of timber ; to which we will add another, for
the more facile perforation and boring of elms, and
other timber to make pipes and aquæducts, and the

excavating of columns, to preserve their shafts from splitting, to which otherwise they are obnoxious.

The frames of both these instruments discover themselves sufficiently to the eye, and therefore will need the less description : There is yet this reformation from those which they use both in Norway, and Switzerland ; that whereas they make the timber approach the saws, by certain indented wheels with a rochet (which is frequently out of order) there is in the first figure, a substitution of two counterpoises of about three hundred pound weight, each as you may see at A. A. fastning the cords to which they append, at the extreams of two movable pieces of timber, which slide on two other pieces of fixed wood, by the aid of certain small pullies, which you may imagine to be within an hinge in the house or mill, by which means the weights continually draw and advance the moving pieces of wood, and consequently the timber to be slit, fastned 'twixt the said pieces, towards the teeth of the saws, rising and falling as the motion of the wheel directs : And on this frame you may put four or five saws, or more if you please, and place them at what intervals you think fit, according to the dimensions which you design in cutting the timber for your use ; and when the piece is sawn, then one or two men with a lever must turn a roller, to which there is annext a strong cord, which will draw back the piece, and lift up the counter-poise ; and so the piece put a little towards one side, direct the saws against another.

The second figure for boring, consists of an ax-tree, to which is fastned a wheel of six and thirty teeth, or more, as the velocity of the water-motion require ; for if it be slow, more teeth are requisite : There

must also be a pinion of six, turn'd by the said indented wheel : Then to the ax-tree of the pinion is to
be fix'd a long auger, as in letter A, which must pass
through the hole B, to be opened, and clos'd as occasion requires, somewhat like a turner's lathe ; the
tree or piece of timber to be bored, is to be plac'd on
the frame CD, so as the frame may easily slide by the
help of certain small wheels, which are in the hollow
of it, and turn upon strong pins, so as the work-man
may shove forwards, or draw the tree back, after 'tis
fastned to the frame ; that so the auger turning the
end of the tree, may be applied to it ; still remembring
to draw it back at every progress of three, or four
inches, which the auger makes for the cleansing it
from the chips, lest the auger break : Continue this
work till the tree, or piece of timber be bored as far
as you think convenient, and when you desire to in-
large the hole, change your auger bits as the figure
represents them.

To these we might add several more, as they are
described by Besson, Ramelli, Cause, and others ; as
likewise cranes and machines for the easier elevation,
moving and transporting of timber, but they are now
become familiar, and therefore I omit them.

Notwithstanding all this, I could wish, that the
most effectual and proper tools for carpentry-work,
and other useful inventions for the raising and transport of great and massive timber, and the like mechanic
uses, were describ'd and explain'd by some persons
expert in the French tongue, and proper English
terms ; together with the figures, as they are publish'd
in *Monsieur* Feliben's *Principles of Architecture*, as of
greater use for our plantations abroad.

The fallen leaves of trees in woods, which lie

sometimes very thick and deep, should be rak'd and shovel'd up, being dry, are very useful for the covering of tender kitchen garden plants, in Winter, instead of litter ; and the rest, if buried in some hole to rot, when dry'd and reduc'd to powder, becomes excellent mould : I wonder this husbandry is so much neglected.

CHAPTER IV.

Of Timber, the Seasoning and Uses, and of Fuel.

Since it is certain and demonstrable, that all arts and artisans whatsoever, must fail and cease, if there were no timber and wood in a nation (for he that shall take his pen, and begin to set down what art, mystery, or trade belonging any way to human life, could be maintain'd and exercis'd without wood, will quickly find that I speak no paradox) I say, when this shall be well consider'd, it will appear, that we had better be without gold, than without timber : This contemplation, and the universal use of that precious material (which yet is not of universal use 'till it be duly prepar'd) has mov'd me to design a solemn chapter for the seasoning, as well as to mention some farther particular application of it. The first, and chiefest use of timber was doubtless for the building of houses and habitations to shelter men in : It is in his 1st. chap. 2. lib. where Vitruvius shews, in what simple, and plain manner, our first progenitors erected their humble cottages ; when like those of Colchis and Phrygia, they began to creep out of the subterranean, and cavernous rocks, and laid the first groundsil upon which they plac'd the upright posts, and rudely fram'd a pointed roof, *arboribus perpetuis planis* (on which the critics have vext their researches) and from which mean beginning, all the superb, and pompous effects of architecture have proceeded : But to pursue our title, we have before spoken concerning some preparations of standing trees design'd for timber, by a half-cutting, disbarking, and the seasons of drawing and using it.

2. Lay up your timber very dry, in an airy place (yet out of the wind or sun) and not standing upright, but lying along one piece upon another, interposing some short blocks between them, to preserve them from a certain mouldiness which they usually contract while they sweat, and which frequently produces a kind of *fungus*, especially if there be any sappy parts remaining.

3. Some there are yet, who keep their timber as moist as they can, by submerging it in water, where they let it imbibe to hinder the cleaving ; and this is good in fir, both for the better stripping and seasoning; yea, and not only in fir, but other timber : Lay therefore your boards a fortnight in the water, (if running the better, as at some mill-pond head) and then setting them upright in the sun and wind, so as it may freely pass through them, (especially during the heats of Summer, which is the time of finishing buildings) turn them daily, and thus treated, even newly sawn boards, will floor far better than a many years dry seasoning, as they call it. But to prevent all possible accidents, when you lay your floors, let the joynts be shot, fitted, and tacked down only for the first year, nailing them for good and all the next ; and by this

KK

means they will lye stanch, close, and without shrinking in the least, as if it were all of one piece ; and upon this occasion I am to add an observation which may prove of no small use to builders ; that if one take up deal-boards that may have lain in the floor an hundred years, and shoot them again, they will certainly shrink (*toties quoties*) without the former method. Amongst wheel-wrights the water-seasoning (which hinders the exhaling of the alcaly salt in it, causing the hardness) is of especial regard, and in such esteem amongst some, that I am assur'd, the Venetians for their Provision in the arsenal, lay their oak some years in it, before they employ it. Indeed the Turks, not only fell at all times of the year, without any regard to the season ; but employ their timber green and unseason'd ; so that though they have excellent oak, it decays in a short time by this only neglect.

Elm fell'd never so green for sudden use, if plung'd four or five days in water, (especially salt) which is best, obtains an admirable seasoning, and may immediately be us'd. I the oftner insist on this water-seasoning, not only as a remedy against the worm, but for its efficacy against warping and distorsions of timber, whether us'd within, or expos'd to the air. Some again commend buryings in the earth, others in wheat ; and there be seasonings of the fire, as for the scorching and hardning of piles, which are to stand either in the water, or the earth.

[1] .... The oak
Explore, suspended in the chimney smoak.

[1] Et suspensa focis explorat robora fumus.
Georg. I.

For that to most timber it contributes much to its duration. Thus do all the elements contribute to the art of seasoning. The learned interpreter of Antonio Neri's *Art of glass*, c. 5. speaking of the difference of vegetables, as they are made use of at various seasons, observes from the button-mould-makers in those woods they use, that pear-trees cut in summer work toughest, but holly in the Winter, box hardest about Easter, but mellow in Summer, hawthorn kindly about October, and service-tree in the Summer.

4. And yet even the greenest timber is sometimes desirable for such as carve and turn ; but it choaks the teeth of our saws ; and for doors, windows, floors, and other close works, it is altogether to be rejected ; especially where walnut-tree is the material, which will be sure to shrink : Therefore it is best to chuse such as is of two or three years seasoning, and that is neither moist nor over-dry ; the mean is best. Sir Hugh Plat informs us that the Venetians use to burn and scorch their timber in a flaming fire, continually turning it round with an engine, till they have gotten upon it an hard, black, coaly crust ; and the secret carries with it great probability ; for that the wood is brought by it to such a hardness and dryness, *ut cum omnis putrefactio incipiat ab humido*, nor earth, nor water can penetrate it ; I my self remembring to have seen charcoals dug out of the ground amongst the ruins of ancient buildings, which have in all probability, lain cover'd with earth above 1500 years.

5. Timber which is cleft, is nothing so obnoxious to rift and cleave as what is hewen ; nor that which is squar'd as what is round ; and therefore where use is to be made of huge and massie columns, let them

be boared through from end to end ; it is an excellent preservative from splitting, and not unphilosophical ; though to cure this accident, the rubbing them over with a wax-cloth is good, painters putty, &c. or before it be converted, the smearing the timber over with cow-dung, which prevents the effects both of sun and air upon it ; if of necessity it must lie expos'd : But besides the former remedies, I find this, for the closing of the chops and clefts of green timber, to anoint and supple it with the fat of powder'd beef-broth, with which it must be well soak'd, the chasms fill'd with spunges dipt into it ; this, to be twice done over : Some carpenters make use of grease and saw-dust mingled ; but the first is so good a way (says my author) that I have seen wind-shock-timber so exquisitely closed, as not to be discerned where the defects were : This must be us'd when the timber is green.

6. We spake before of squaring, and I would now recommend the quartering of such trees as will allow useful and competent scantlings, to be of much more durableness and effect for strength, than where (as custom is, and for want of observation) whole beams and timbers are apply'd in ships or houses, with slab and all about them, upon false suppositions of strength beyond these quarters : For there is in all trees an evident interstice or separation between the heart and the rest of the body, which renders it much more obnoxious to decay and miscarry, than when they are treated and converted as I have describ'd it ; and it would likewise save a world of materials in the building of great ships, where so much excellent timber is hew'd away to spoil, were it more in practice. Finally,

7. I must not omit to take notice of the coating of timber in work, us'd by the Hollanders for the preservation of their gates, port-cullis's, draw-bridges, sluces, and other huge beams and contignations of timber expos'd to the sun, and perpetual injuries of the weather, by a certain mixture of pitch and tar, upon which they strew small pieces of cockle, and other shells, beaten almost to powder, and mingled with sea-sand, or the scales of iron, beaten small and sifted, which incrusts, and arms it after an incredible manner against all these assaults and foreign invaders : But if this should be deem'd more obnoxious to firing, I have heard that a wash made of allum has wonderfully protected it against the assaults even of that devouring element, and that so a wooden tower or fort at the Piræum an Athenian port, was defended by Archelaus a commander of Mithridates, from the great Sylla : But you have several compositions for this purpose in that incomparable treatise of naval architecture, written in the Low-Dutch, by N. Witsen, chap. 6. part. 1. the book is a folio, and he that should well translate it into our language (which I much wonder has not yet been done) would deserve well of the publick.

8. Timber that you have occasion to lay in mortar, or which is in any part contiguous to lime, as doors, window-cases, groundsils, and the extremities of beams, &c. have sometimes been capp'd with molten pitch, as a marvelous preserver of it from the burning and destructive effects of the lime ; but it has since been found rather to heat and decay them, by hindring the transudation which those parts require ; better supply'd with loam or strowings of brick-dust, or pieces of boards ; some leave a small hole for the

air. But though lime be so destructive whilst timber
lies thus dry, it seems they mingle it with hair, to
keep the worm out of ships which they sheath for
southern voyages ; though it is held much to retard
their course : Wherefore the Portugals scorch them
with fire, which often proves very dangerous ; and
indeed their timber being harder, is not so easily
penetrable ; and therefore have some been thinking
of finding out some tougher sorts of materials, especially of a bitter sap ; such as is reported to be the
wood of a certain Indian-pear : And some talk of a
lixivium to do the feat ; others of a pitchy substance
to be extracted out of sea-coal; but nothing has yet
been found more expedient, than the late application
of thin lamins of sheet-lead, if that also be no impedi-
ment to their sailing : However, there are many
kind of woods in the Western-Indies (besides the
acajou) that breed no worms, and such is the white
wood of Jamaica, proper enough to build ships. In
the mean time, let me not omit what the learned
Dr. Lister in his *Notes* upon Godariius of *Insects*, says,
that he is persuaded there could not be a more probable expedient to discover what kind of timber were
best for sheathing, than to tye certain polished pieces
of wood (cut like tallies) to a buoy, in some waters
and streams much infested with the worms ; for that
sort of wood which the worm should refuse, would
in all reason be chosen for the use desir'd. The
Indies being stor'd with greater varieties of timber
than Europe, it were probable there might some be
found, which that kind of river-worm will never
attack.

9. For all uses, that timber is esteem'd the best,
which is the most ponderous, and which lying long

makes deepest impression in the earth, or in the water being floated ; also what is without knots, yet firm, and free from sap ; which is that fatty, whiter, and softer part, call'd by the Ancients *alburnum*, which you are diligently to hew away ; here we have much ado about the *torulus* of the fir, and the Φλοιώδης κύκλος by both Vitruvius and Theophrastus, which I pass over. You shall perceive some which has a spiral convolution of the veins ; but it is a vice proceeding from the severity of unseasonable Winters, and defect of good nutriment.

10. My Lord Bacon, *Exp.* 658. recommends for tryal of a sound or knotty piece of timber, to cause one to speak at one of the extreams to his companion listning at the other ; for if it be knotty, the sound (says he) will come abrupt.

11. Moreover, it is expedient that you know which is the grain, and which are the veins in timber, (whence the term *fluviari arborem*) because of the difficulty of working against it : Those therefore are counted the veins which grow largest, and are softer for the benefit of cleaving and hewing ; that the grain or pectines, which runs in waves, and makes the divers and beautiful chamfers which some woods abound in to admiration. The fir-tree horizontally cut, has two circles of different fibres, which (when the timber comes to be cleft in the middle) separates into four different waves, whence Pliny calls them *quadrifluvios*, and it is to be noted, that the nodous, and knotty part of these sort of trees, is that only which grows from the first boughs to the summit or top, by Vitruvius term'd the *fusterna*, which both Baldus, and Salmasius derive *a fuste*. The other clean part, free of these boils, (being that which when the

sappy slab is cut away, is the best) he calls *sapiena*.
Finally, the grain of beech runs two contrary ways,
and is therefore to be wrought accordingly ; and
indeed the grain of all timber ought well to be
observ'd ; since the more you work according to it,
especially in cleaving, and the less you saw, the
stronger will be your work.

12. Here it may be fitly enquir'd, whether of all
the sorts we have enumerated, the old, or the younger
trees do yield the fairest colour, pleasant grain and
gloss for wainscot, cabinets, boxes, gunstocks, &c. and
what kind of pear and plumtree give the deepest red,
and approaches nearest in beauty to Brasil : 'Tis
affirm'd the old oak, old walnut, and young ash, are
best for most uses, and yet for ship-carpentry this
does not always hold ; nor does the bigness of it so
much recommend it ; because 'tis commonly a sign
of age, which (like to very old men) is often brittle
and effete. Black and thorny plum-tree is of the
deepest oriency ; but whether these belong to the
forest, I am not yet satisfied, and therefore have
assigned them no chapter apart. But now I speak
of the plum-tree, I am assur'd by a worthy friend,
that the gum thereof dissolv'd in vinegar, does cure
the most contumacious tetters, when all other remedies
outward or inwardly applied, nothing avail'd.

13. Lastly, I would also add something concerning
what woods are observed to be most sonorous for
musical instruments : We as yet detect few but the
German *aer* which is a species of maple, for the
rimms of viols, and the choicest and finest grain'd fir
for the bellies : The finger-boards, back, and ribs,
I have seen of yew, pear-tree, &c. but pipes, recorders,
and wind-instruments, are made both of hard, and

soft woods; I had lately an organ with a sett of oaken-pipes, which were the most sweet and mellow that were ever heard; It was a very old instrument, and formerly, I think, belonging to the Duke of Norfolk. We shall say nothing of the other various uses of timber superstitiously mention'd, when we find they might not carve the statues of the pagan gods of every sort of wood, *ne quovis ex ligno fiat Mercurius*; but of this by the way.

14. For the place of growth, that timber is esteem'd best which grows most in the sun, and on a dry and hale ground; for those trees which suck, and drink little, are most hard, robust, and longest liv'd, instances of sobriety. The climate contributes much to its quality, and the northern situation is preferred to the rest of the quarters; so as that which grew in Tuscany was of old thought better, than that of the Venetian side; and yet the Biscay timber is esteemed better than what they have from colder countries: And trees of the wilder kind, and barren, than the over-much cultivated, and great bearers: But of this already.

15. To omit nothing, authors have summ'd up the natures of timber; as the hardest ebony, box, larch, lotus, terebinth, *cornus*, yew, &c. and though these indurated woods be too ponderous for ship-carpentry; yet there have been vessels built of them by the Portuguezes in America; in which the planks, and innermost timbers had been saw'd very thin for lightness sake, and the knee-timber put together of divers small pieces, by reason of the inflexibleness of it, both which could not but render the ships very weak: In the mean time, the perfection of these hard materials consists much in their receiving the most

exquisite politure ; and for this, lin-seed, or the sweeter nut-oyl does the effect best : Pliny gives us the receipt, with a decoction of walnut-shells, and certain wild pears : Next to these, oak, for ships, and houses (or more minutely) the oak for the keel, the *robur* for the prow, walnut the stern, elm the pump ; Furnerus l. 1. c. 22. conceives the ark to have been built of several woods ; cornell, holly, &c. for pins, wedges, &c. chesnut, horn-beam, poplar, &c. then for bucklars, and targets, were commended the more soft and moist ; because apt to close, swell, and make up their wounds again ; such as willow, lime, birch, alder, elder, ash, poplar, &c.

The *robur*, or wild-oak-timber, best to stand in ground ; the *quercus* without ; and our English, for being least obnoxious to splinter, and the Irish for resisting the worm (tough as leather) are doubtless for shipping to be preferr'd before all other : The cypress, fir, pines, cedar, &c. are best for posts, and columns, because of their erect growth, natural and comely diminutions. Then again it is noted, that Oriental trees are hardest towards the *cortex* or bark, our Western towards the middle which we call the heart ; and that trees which bear no fruit, or but little, are more durable than the more pregnant. It is noted of oak, that the knot of an inveterate tree, just where a lusty arm joins to the stem, is as curiously vein'd as the walnut, which omitted in the chapter of the oak, I here observe. The *palmeto* growing to that prodigious height in the Barbadoes, and whose top bears an excellently tasted cabage, grows so wonderfully hard, that an edge-tool will scarce be forced into it.

Pines, pitch, alder, and elm, are excellent to make

pumps and conduit-pipes, and for all water-works, &c. fir for beams, bolts, bars ; being tough, and not so apt to break as the hardest oak : In sum, the more oderiferous trees are the more durable and lasting ; and yet I conceive that well- season'd oak may contend with any of them ; especially, if either preserved under ground, or kept perfectly dry ; in the mean time, as to its application in shipping, the best of it ought to be employ'd for the keel, (that is, within, else elm exceeds) the main beams and rafters, whilst for the ornamental parts, much slighter timber serves : One note more is requisite, namely, that great care be had to make the trundels of the best, toughest, and sincerest part, many a vessel having been lost upon this account ; and therefore dry and young timber is to be preferr'd for this, and for which the Hollanders are plentifully furnish'd out of Ireland, as Nicholas Witsen has himself acknowledged.

Is it not after all this to be deplor'd, that we who have such perpetual use and convenience for ship-timber, should be driven to procure it of foreign stores, so many thousand loads, at intolerable prices : But this we are oblig'd to do and supply from the Eastern countries, as far as Norway, Poland, Prussia, Dantzick, and farther, even from Bohemia, tho' greatly impair'd by sobbing so long in the passage : But of this the most industrious, and our worthy friend Mr. Pepys, (late Secretary of the Admiralty) has given a just and profitable account in his *Memoirs*.

16. Here farther for the uses of timber, I will observe to our reader some other particulars for direction both of the seller and buyer, applicable to the several species : And first of the two sorts of lathes allow'd by statute, one of five, the other of four foot

long, because of the different intervals of rafters :
That of five has 100 to the bundle, those of four 120;
and to be in breadth 1 inch and $\frac{1}{2}$, and half inch thick;
of either of which sorts there are three, *viz.* heart-
oak, sap-lathes, and deal-lathes, which also differ
in price : The heart-oak are fittest to lie under
tyling ; the second sort, for plastring of side-walls,
and the third for ceilings, because they are streight
and even.

17. Here we will gratifie our curious reader with
as curious an account of the comparative strength and
fortitude of the several usual sorts of timber, as upon
suggestions previous to this work, it was several times
experimented by the Royal Society, tho' omitted in
the impression, because the tryals were not complete
as they now thus stand in our Register.

MARCH 23. 1663.

The experiment of breaking several sorts of wood
was begun to be made : And there were taken three
pieces of several kinds ; of fir, oak, and ash, each an
inch thick, and two foot long, the fir weighed $8\frac{1}{8}$
ounces, and was broken with 200 l. weight : The
oak weighed $12\frac{3}{4}$ ounces, broken with 250 weight :
The ash weigh'd $10\frac{1}{4}$ ounces, broken with 325 weight.

Besides there were taken 3 pieces of the same sort
of wood, each of $\frac{1}{2}$ inch thick, and 1 foot long : The
fir weigh'd 1 ounce, and was broken with $\frac{5}{8}$ of an 100:
The oak weigh'd $1\frac{5}{8}$ ounces, broken with $\frac{5}{8}$ of an 100 :
The ash weighed $1\frac{3}{8}$ ounces, broken with 100 l.

Again, there was a piece of fir $\frac{1}{2}$ inch square, and
two foot long, broken with 33 l. A piece of $\frac{1}{2}$ inch
thick, 1 inch broad, and 7 foot long, broken with

100 weight edge-wise ; and a piece of $\frac{1}{2}$ inch thick, 1$\frac{1}{2}$ broad, 2 foot long, broken with 125 weight, also edge-wise.

The experiment was order'd to be repeated and recommended by the President, to Sir Will. Petty and Dr. Hook ; and it was suggested by some of the company, that in these tryals consideration might be had of the age, knottiness, solidity, several soils and parts of trees, &c. and Sir Robert Morray did particularly add, that it might be observed how far any kind of wood bends before it breaks.

MARCH .... 64.

The operator gave an account of more pieces of wood broken by weight, *viz.* a piece of fir 4 foot long 2 inches, 53 ounce weight, broken with 800 l. weight, and very little bending, with 750 ; by which the hypothesis seems to be confirm'd, that in similar pieces, the proportion of the breaking-weight is according to the basis of the wood broken : Secondly of a piece of fir two foot long, one inch square, cut away from the middle both ways to half an inch, which supported 250 l. weight before it broke, which is more by 50 l. than a piece of the same thickness every way was formerly broken with ; the difference was guessed to proceed from the more firmness of this other piece.

His Lordship the President, was desired to contribute to the prosecution of this experiment, and particularly, to consider what line a beam must be cut in, and how thick it ought to be at the extream, to be equally strong : Which was brought in April 13, but I find it not enter'd.

APRIL 20. 1664.

The experiment of breaking wood was prosecuted, and there were taken two pieces of fir, each two foot long, and 1 inch square, which were broken, the one long-ways with 300 l. weight, the other transverse-ways with 2½ hundred : Secondly, two pieces of the same wood, each of ¾ of an inch square, and two foot long, broken, the one long-ways with 1¼ hundred ; the other transverse, with 100 l. weight : Thirdly, one piece of two foot long ½ inch square, broken long-ways with 81 l. Fourthly, one piece cut out of a crooked oaken-billet, with an arching grain, about ¾ inch square, two foot long, broken with ¾ hundred.

JUNE 29. 1664.

There were made several experiments more of breaking wood : First, a piece of fir, ½ inch diameter and 3 inches long, at which distance the weight hung, broke in the plane of the grain horizontally, with 66¾ l. whereof 15 l. *Troy* ; vertically, with 2 l. more.   Also fir of ¼ inch diameter, and 1½ inch long, broke vertically with 20 l. and horizontally, with 19 l. Elm of ½ inch diameter, and three inches long, broke horizontally, with 47 l. vertically with 23 l. Elm of ¼ inch diameter, and 1½ inch long, broke horizontally with 12 l. vertically with 10 l. which is note-worthy.

JULY 6. 1664.

The experiment of breaking woods prosecuted : A piece of oak of ½ inch diameter, and three inches long,

at which distance the weight hung, broke horizontally
with 48 l., vertically with 40 l.; ash of ½ inch diameter,
and 3 inch long, horizontally with 77 l., vertically,
with 75 l.; ash of ½ inch diameter, and 1½ inch long,
horizontally with 19 l. vertically, with 12 l. &c.
Thus far the Register.

In the mean time I learn, that in the mines of
Mendip, pieces of timber, of but the thickness of a
man's arm, will support ten tun of earth ; and that
some of it has lain 200 years, which is yet as firm as
ever, growing tough and black, and being expos'd
two or three days to the wind and sun, scarce yields
to the ax.

18. Here might come in the problems of Cardinal
Cusanus in lib. 4. *Idiotae dial.*4to, concerning the different velocity of the ascent of great pieces of timber,
before the smaller, submerged in water ; as also of the
weight; as *e. g.*   Why a piece of wood 100 l. weight,
poising more in the air than 2 l. of lead, the 2 l. of
lead should seem to weigh (he should say sink) more
in [1] the water ?   Why fruits being cut off from the
tree, weigh heavier, than when they were growing ?
with several the like paradoxes, haply more curious
than useful, and therefore we purposely omit them ;
but so may we not the recommendation of that useful
treatise of duplicate proportion, together with a new
hypothesis of elastique or springy bodies, to shew the
strengths of timbers, and other homogeneous materials
apply'd to buildings, machines, &c. as it is published by
that admirable genius, the learned Sir William Petty.
To which we join that part of Dr. Grew's comparative
Anatomy of trunks, as variously fitted for mechanical

---

[1] Of the specific gravity of timber in proportion to water, see the table in *Philos. Transact.* n. 169, and 199.

uses; where that most industrious and curious searcher into nature, describes to us whence woods are soft, fast, hard, apt to be cleft, tough, durable, &c. Lastly,

19. Concerning squar'd and principal timber, for any usual buildings, these are the legal proportions, and which buildings ought not to vary from.

Summers or girders from	F. 14, 18 to 20, 20, 23, 26	F. 16, 20, 23, 26, 28	in length, must be in their square,	In. 11, 13 &, 14, 16, 17	In. 8, 9, 10, 12, 17	Joysts of	11½ / 10½	in length must be in their square	In. In. 8 — 3 / 7 & 3 / 6 — 3

Binding joysts and trimmers from	F. F. 7 to 11½	in length must be in their square	6 / 7 & / 8	5 / 5 / 5	Wall-plates and beams of any length, from 15 foot, may have in their square........	In. 7 / 10 & / 8	In. 5 / 6 / 6

Purlynes from	F. 15 to 18½ / 18	F. 18½ / 21½	in length, must have in their square	9 — 8 & 12 — 9

Principal rafters cut taper from	F. 12½—14½, 14½—18¼, 18½ to 21½, 21½—24½, 24½—26½	F.	in length must have in their square on one side	In. In. 8 5 / 9 7 / 10 to 8 / 12 9 / 9 9	on the other side	6 / 7 / 8 / 8 / 9	Single rafters in length from 6½ to 9½	F. 6½ / 0 / 9½	must have in their squ.	5 -3½ & 5 - 4

Principal dischargers of any length from	Foot 10 upward	must have in their square	In. In. 13 — 12 / 16 — 13

But carpenters also work by square, which is 10 foot in framing and erecting the carcase (as they call it) of any timber edifice, which is valued according to the goodness and choice of the materials, and curiosity in framing; especially roofs and stair-cases, which are of most charges. And here might also something be added concerning the manner of framing the carcases of buildings, as of floors, pitch of roofs, the length of hips and sleepers, together with the names of all those several timbers used in fabricks, totally consisting of

wood ; but I find it done to my hand, and publish'd
some years since, at the end of a late translation of the
first book of Palladio, to which I refer the reader.
And to accomplish our artist in timber, with the
utmost which that material is capable of ; to the study
and contemplation of that stupendious roof, which
now lies over the ever renowned Sheldonean Theatre
at the University of Oxford ; being the sole work and
contrivement of my most honoured friend, Sir Christopher Wren, now worthily dignified with the superin-
tendency of the Royal buildings.   See Dr. Plot's
description of it in his *Nat. Hist. of Oxfordshire*, 272,
273. tab. 13, 14. also Dr. Wallis *de Motu*, part 3. *de
Vecte*, cap. 6. prop. 10.

   Other conversions there are of timber of all lengths,
sizes and dimensions, for arches, bridges, floors and
flat-work, (without the supports of pillars) tables,
cabinets, inlayings and carvings, skrews, &c. with the
art of turning ; to the height of which divers gentlemen have arriv'd, and for their diversion, produc'd
pieces of admirable invention and curiosity : These, I
say, belonging to the mechanick uses of timber, might
enter here ; with a catalogue of innumerable models
and other rarities, (to be found in the repositories and
collections of the curious.)   But let this suffice.

   20. We did, in chap. 21. mention certain subterranean trees, which Mr. Camden supposes grew
altogether under the ground : And truly it did appear
a very paradox to me, till I both saw, and diligently
examin'd that piece (plank, stone, or both shall I
name it ?) of *lignum fossile* taken out of a certain
quarry thereof at Aqua Sparta, not far from Rome,
and sent to the most incomparably learned Sir George
Ent, by that obliging *virtuoso* Cavalier dal. Pozzo.

He that shall examine the hardness, and feel the ponderousness of it, sinking in water, &c. will easily take it for a stone ; but he that shall behold its grain, so exquisitely undulated, and varied, together with its colour, manner of hewing, chips, and other most perfect resemblances, will never scruple to pronounce it arrant wood.

Signior Stelluti (an Italian) has publish'd a whole treatise expresly to describe this great curiosity : And there has been brought to our notice, a certain relation of an elm growing in Bark-shire, near Farringdon, which being cut towards the root, was there plainly petrified ; the like, as I once my self remember to have seen in another tree, which grew quite through a rock near the sepulchre of Agrippina (the mother of that monster Nero) at the Baia by Naples, which appear'd to be all stone, and trickling down in drops of water, if I forget not. But, whilst others have philosophiz'd according to their manner upon these extraordinary concretions, see what the most industrious and knowing Dr. Hook, Curator of this Royal Society, has with no less reason, but more succinctness, observ'd from a late microscopical *examen* of another piece of petrify'd wood ; the description and ingenuity whereof cannot but gratifie the curious, who will by this instance, not only be instructed how to make enquiries upon the like occasions ; but see also with what accurateness the Society constantly proceeds in all their indagations, and experiments ; and with what candour they relate, and communicate them.

21. It resembled wood, in that

' First, all the parts of the petrify'd substance ' seem'd not at all dislocated or alter'd from their ' natural position whiles they were wood ; but the

' whole piece retain'd the exact shape of wood, having
' many of the conspicuous pores of wood still re-
' maining, and shewing a manifest difference visible
' enough between the grain of the wood and that of
' the bark ; especially, when any side of it was cut
' smooth and polite; for then it appear'd to have a very
' lovely grain, like that of some curious close wood.

' Next (it resembled wood) in that all the smaller,
' and (if so I may call those which are only to be
' seen by a good glass) microscopical pores of it, appear
' (both when the substance is cut and polish'd trans-
' versly, and parallel to the pores) perfectly like the
' microscopical pores of several kinds of wood, re-
' taining both the shape and position of such pores.

' It was differing from wood,

' First, in weight, being to common water, as 3 $\frac{1}{4}$
' to 1. whereas there are few of our English woods
' that, when dry, are found to be full as heavy as water.

' Secondly, in hardness, being very near as hard as
' flint, and in some places of it also resembling the
' grain of a flint ; it would very readily cut glass, and
' would not without difficulty (especially in some
' parts of it) be scratch'd by a black hard flint : it
' would also as readily strike fire against a steel, as
' also against a flint.

' Thirdly, in the closeness of it ; for, though all the
' microscopical pores of the wood were very conspic-
' uous in one position, yet by altering that position of
' the polish'd surface to the light, it also was manifest
' that those pores appear'd darker than the rest of the
' body, only because they were fill'd up with a more
' dusky substance, and not because they were hollow.

' Fourthly, in that it would not burn in the fire ;
' nay, though I kept it a good while red-hot in the

' flame of a lamp, very intensely cast on it by a blast
' through a small pipe ; yet it seemed not at all to
' have diminish'd its extension ; but only I found it to
' have chang'd its colour, and to have put on a more
' dark and dusky brown hue.   Nor could I perceive
' that those parts which seem'd to have been wood at
' first, were any thing wasted, but the parts appear'd
' as solid and close as before.   It was farther observ-
' able also, that as it did not consume like wood, so
' neither did it crack and fly like a flint, or such like
' hard stone ; nor was it long before it appear'd red-hot.

  ' Fifthly, in its dissolubleness ; for putting some
' drops of distilled vinegar upon the stone, I found it
' presently to yield very many bubbles, just like those
' which may be observed in spirit of vinegar when it
' corrodes coral ; tho' I guess many of those bubbles
' proceeded from the small parcels of air, which were
' driven out of the pores of this petrify'd substance,
' by the insinuating liquid *menstruum*.

  ' Sixthly, in it's rigidness, and friability ; being not
' at all flexible, but brittle like a flint ; insomuch, that
' with one knock of a hammer I broke off a small
' piece of it, and with the same hammer quickly beat
' it to pretty fine powder upon an anvil.

  ' Seventhly, it seem'd also very differing from wood
' to the touch, feeling more cold than wood usually
' does, and much like other close stones and minerals.

  ' The reason of all which phaenomena seems to be,
  ' That this petrified wood having lain in some
' place where it was well soaked with petrifying
' water (that is, such a water as is well impregnated
' with stony and earthy particles) did by degrees se-
' parate, by straining and filtration, or perhaps by pre-
' cipitation, cohesion or coagulation, abundance of

' stony particles from that permeating water : Which
' stony particles having, by means of the fluid vehicle,
' convey'd themselves not only into the microscop-
' ical pores, and perfectly stopp'd up them, but also
' into the pores, which may perhaps be even in that
' part of the wood which through the microscope
' appears most solid ; do thereby so augment the
' weight of the wood, as to make it above three times
' heavier than water, and perhaps six times as heavy
' as it was when wood : Next, they hereby so lock
' up and fetter the parts of the wood, that the fire
' cannot easily make them fly away, but the action of
' the fire upon them is only able to char those parts as
' it were, like as a piece of wood if it be closed very
' fast up in clay, and kept a good while red-hot in
' the fire, will by the heat of the fire be char'd, and
' not consum'd ; which may perhaps be the reason
' why the petrify'd substance appear'd of a blackish
' brown colour after it had been burnt. By this in-
' trusion of the petrify'd particles it also becomes hard,
' and friable ; for the smaller pores of the wood
' being perfectly stuffed up with these stony particles,
' the particles of the wood have few or no pores in
' which they can reside, and consequently, no flexion
' or yielding can be caus'd in such a substance. The
' remaining particles likewise of the wood among the
' stony particles may keep them from cracking and
' flying, as they do in a flint.

The casual finding of subterraneous trees has been
the occasion of this curious digression, besides what
we have already said in cap. III. book II. Now it
were a strange paradox to affirm, that the timber
under the ground, should to a great degree, equal the
value of that which grows above the ground ; seeing

though it be far less, yet it is far richer ; the roots of the vilest shrub being better for its toughness, and for ornament, and delicate uses, much more preferrable than the heart of the fairest and soundest tree : And many hills, and other waste-places, that have in late and former ages been stately groves and woods, have yet this treasure remaining, and perchance sound and unperish'd, and commonly (as we observ'd) an hindrance to other plantations ; engines therefore, and expedients for the more easily extracting these cumbrances, and making riddance upon such occasions, besides those we have produc'd, would be excogitated and enquir'd after, for the dispatch of this difficult work.

Thus from all these instances, we may gather the necessity of a more than ordinary knowledge, requisite in such whose profession obliges them that deal in timber, to study the art well ; nor is it a small stock of philosophy, to skill in the nature and property of these materials, and which does not only concern architects, but their subsidiary, carpenters, joyners, especially wood-brokers, &c. I cannot therefore but take notice, that among the ancient *sportulae*, bequeath'd by several founders and foundresses, to incourage the gardiners, *dies violaris*, and *rosalis*, (which was about the time of the *Floraria*) there was among the Romans a College or Hall, not unlike that of our carpenters ; where, upon a certain day, the fraternity not only met to feast, but doubtless to confer and edify one another ; as appears by an ancient inscription of the *Dendrophori* at Puteoli, mention'd by the learned [1] Dr. Spon, which for the honour of our present discourse we subjoin.

[1] *Dissert.* 23. & *Miscellan. Antiq.* Sect. II. *Art.* XI.

EX. S. C. DENDROPHORI. CREATI. QVI.
SUNT. SUB. CURA. XV. VIR. ST. CC. V. V.
PATRON. L. AMPIUS. STEPHANUS. SAC. M.
DEI. Q. Q. DEDICATIONI. HUJUS. PANEM.
VINUM. ET. SPORTULAS. DEDIT. HERCU-
LANUS.
 C. VALERIUS. PICENT. VI. C. JULIUS.
  LONGINIUS JUSTINUS.

With all the rest (a numerous catalogue) of the consuls names ; it being it seems, a corporation establish'd by the state, when they carried boughs and
branches of trees in procession, and distributed a
sportula of bread and wine : But of this, and of the
fabri, tignarii, naupegiarii, (ship-carpenters) and *centon-
arii,* see this learned man's excellent dissertation.

 These Colleges or Halls were dedicated to Diana,
as goddess of the woods ; of which another Roman
inscription is yet exstant.

D I A N A E.
COLLEG. NAUPEGIAR.
M. JUNIUS. BALISTUS.
ET. Q. AVILLIUS. EROS.
II. VIR. D. D.

 23. Finally, for the use of our chimnies, and main-
tenance of fire, the plenty of wood for fuel, rather
than the quality is to be looked after ; and yet there
are some greatly to be preferr'd before others, as
harder, longer-lasting, better heating, and chearfully

---

[1] The Jews had their Feast of Ξυλοφορία, mention'd by Josephus, in which they
were oblig'd to carry wood to the Temple for the maintaining the fires of the
altar.

burning ; for which we have commended the ash,
&c. in the foregoing paragraphs, and to which I pretend not here to add much, for the avoiding repetitions ; though even an history of the best way of
charring would not mis-become this discourse.

But something more is to be said sure, concerning
the felling of *lignum*, fuel-wood, (for so critics would
distinguish it from *materia*, timber :) Benedictus Cursius, *Hortor*. L. VIII. C. XI. reckons up what woods
make the best firing ; also of coaling & *de facibus*,
clearing, and what else belongs to ξυλοτομία, especially
for the use of [1] sacrifices, which had their particular
sorts ; as in the temple of Despoena in Arcadia, where
they were prohibited the burning of olive-wood, or
the φυτὸν μανικόν, the vaticinatric laurel, or the thick-
rin'd oak, nor any *fungus* or rotten wood, but what
was well dry'd, and apt to kindle without smoaking.
In the sacrifice of Jupiter they us'd white poplar, the
pine on the altar of Ceres : The Persian *Magi* burnt
their sacrifices with myrtil and the boughs of laurel ;
and in general, all the pagan Gods, that wood which
was sacred to the particular deity : Of all which to
particularize, let the curious enquire. We proceed
therefore with what concerns this most useful chapter.

And first, that our fuelist begin with the under-
wood : Some conceive between Martlemas and Holy-
rood ; but generally with oak, as soon as 'twill strip,
but not after May; and for ashes, 'twixt Michaelmas
and Candlemas; and so fell'd, as that the cattle may
have the browsing of it, for in Winter they will not
only eat the tender twigs, but even the very moss ;
but fell no more in a day than they can eat for this
purpose. This done, kid or bavin them, and pitch

---

[1] v. Eustath. in Odyss. 3.

them upon their ends to preserve them from rotting: Thus the under-wood being dispos'd of, the rest will prosper the better; and besides, it otherwise does but rot upon the earth, and destroy that which would spring. If you head, or top for the fire, 'tis not amiss to begin three or four foot above the timber, if it be considerable; but in case they are only shaken-trees and hedgerows, strip them even to thirty foot high, because they are usually full of boughs: and 'twere good to top such as you perceive to wither at the tops a competent way beneath, to prevent their sickness downwards, which will else certainly ensue ; whereas by this means even dying trees may be preserved many years to good emolument, tho' they never advance taller ; and being thus frequently shred, they will produce more than if suffered to stand and decay : This is a profitable note for such as have old, doating, or any ways infirm woods: In other fellings, some advise never to commence the disbranching from the top, for though the incumbency of the very boughs upon the next, cause them to fall off the easier, yet it endangers the splicing of the next, which is very prejudicial, and therefore advise the beginning at the nearest.   And in cutting for fuel you may as at the top, so at the sides, cut a foot, or more from the body; but never when you shred timber-trees : We have said how dangerous it is, to cut for firewood when the sap is up, it is a mark of improvident husbands ; besides it will never burn well, though abundance be congested : Lastly, remember that east and north-winds are unkind to the succeeding shoots.

Now for directions in stacking (of which we have said something in Chap. of Copp'ces) ever set the lowest course an end, the second that on the sides and ends, viz. sides and ends outward ; the third thwart the other on the side, and so the rest, till all are placed, spending the up-most first.

Thus we have endeavoured to prescribe the best directions we could learn concerning this necessary subject. And in this penury of that dear commodity, and to incite all ingenious persons, studious of the benefit of their country, to think of ways how our woods may be preserved, by all manner of arts which may prolong the lasting of our fuel, I would give the best encouragements. Those that shall seriously consider the intolerable misery of the poor Chauci (the then inhabitants of the Low-Countries) describ'd by Pliny, lib. 16. cap. 1. (how opulent soever their late industry has render'd them) for want only of wood for fuel, will have reason to deplore the excessive decay of our former store of that useful commodity ; and by what shifts our neighbours the Hollanders, do yet repair that defect, be invited to exercise their ingenuity : The process of which is casting the die or square of the turf in 4 equal quarters ; and to build them so up, (as our brick-makers do their crude ware) that they may have the free intercourse of the air till they are dry : See Quicciardius in his description of Holland, or du Cange's *Glossary*, *verbo Turba* : But besides the [1] dung of beasts, and the peat and turf (which we may find in our ouzy lands and heathy commons) for their chimneys, cowsheads, &c. they make use of stoves both portable and standing ; and truly the more frequent use of those inventions in our great wasting cities (as the custom is through all Germany) as also

---

[1] In many places (where fuel is scarce) poor people spread fern and straw in the ways and paths where cattle dung and tread, and then clap it against a wall till it be dry : But that of hogs is very noysome.

of those new and excellent ovens invented by Dr. Kef-
fler, for the incomparably baking of bread, &c. would
be an extraordinary expedient of husbanding our fuel,
as well as the right mingling, and making up of
charcoal-dust and loam, as 'tis hinted to us by Sir
Hugh Plat, and is generally us'd in Maestricht, Liege,
and the country about it ; than which there is not a
more sweet, lasting, and beautiful fuel : The manner
of it is thus :

24. Take about one third part of the smallest of
any coal, pit, sea, or char-coal, and commix them very
well with loam (whereof there is in some places to
be found a sort somewhat more combustible) make
these up into balls (moistned with a little urine of
man or beast) as big as an ordinary goose egg, or
somewhat bigger ; or if you will in any other form,
like brick-bats, &c. expose these in the air till they
are throughly dry ; they will be built into the most
orderly fires you can imagine, burn very clear, give a
wonderful heat, and continue a very long time.   But
first you must make the fire of char-coal or small-coal,
covering them with your eggs, hotshots, or hovilles
(as they are call'd) and building them up in pyramids,
or what shape you please, they will continue a glowing, solemn and constant fire for seven or eight hours
without being stirred, and then they encourage and
recruit the innermost with a few fresh eggs, and turn
the rest, which are not yet quite reduc'd to cinders ;
and this mixture is devis'd to slacken the impetuous
devouring of the fire, and to keep the coals from con-
suming too fast.

Two or three short billets cover'd with char-coal
last much longer, and with more life than twice the
quantity by it self, whether char-coal alone, or billet;

and the billets under the char-coal being undisturb'd, will melt as it were into char-coals of such a lasting size.

If small-coals be spread over the char-coal, where you burn it alone, 'twill bind it to longer continuance ; and yet more, if the small-coal be made of the roots of thorns, briers, and brambles. Consult L. Bacon, *Exp. 775.*

25. The *quercus marina*, wrack, or sea-weed which comes in our oyster-barrels, laid under New-Castle-coal to kindle it (as the use is in some places) will (as I am inform'd) make it out-last two great fires of simple coals, and maintain a glowing luculent heat without waste.   This sort of fuel is much made use of in Malta and the islands thereabout, especially to burn in their ovens, and the peasant who first brought it into custom, I find highly commended by an author as a great benefactor to his country : The manner of gathering it is to cut it in Summer time from the rocks, whereon it grows abundantly, and bringing it in boats or otherwise to land, spread and dry it in the sun like hay, turning and cocking it till it be fully cured : It makes an excellent fire alone, and roasts to admiration ; and when all is burnt, the ashes are one of the best manures for land in the world, for the time it continues in vertue, which should be frequently supplied with fresh ; and as to the fire mingled with other combustibles, it is evident that it adds much life, continuance and aid, to our sullen sea-coal fuel ; and if the main ocean should afford fuel (as the bernacles and soland-geese are said to do in some parts of Scotland, with the very sticks of their nests) we in these isles may thank our selves if we be not warm : These few particulars I have but mention'd to animate improvements, and ingenious

attempts of detecting more cheap and useful processes, for ways of charing-coals, peat, and the like fuliginous materials; as the accomplish'd Mr. Boyl has intimated to us in the fifth of those his precious *Essays* concerning the usefulness of natural philosophy, part II. cap. 7, &c. to which I refer the curious.   In the mean time, were not he worthy a statue of gold, that (*salvo* to our New-Castle-trade and seminary of mariners) should in this penury, and of fire-wood, about so monstrous a devourer, as this vast city (poyson'd with smoak and soot) find out an expedient, that should within the space of five and twenty years, not only free it from all this hellish and pernicious fog, by furnishing it with fuel sufficient to feed and maintain all its hearths and fires with sweet and wholsome billet?   This, the ingenious Mr. Nourse seems to demonstrate, and I think not impossible, whilst my *Fumifugium* is long since vanished *in aura*. There is no very great store of wood about Madrid, where the Winters are sharp and so very piercing, that there is spent no less than four millions of *arrobas* of char-coal (every *arroba* being 3 quarters of our bushel) and pays to the king a real *per arroba* before it comes into the town, or is sold : It is charr'd of the *enzina* or cork-tree ; besides which they use very little fuel-wood, it being exceeding hard, and consequently lasting and sweet.   But to return to the law.

26. By the preamble of the Statute 7 Ed. 6. one may perceive (the measures compar'd) how plentiful fuel was in the time of Ed. the 4th. to what it was in the reigns of his successors : This suggested a review of sizes, and a reformation of abuses ; in which it was enacted, that every sack of coals should contain four bushels ; every taleshide to be four foot

long, besides the carf ; and if nam'd of one, marked one, to contain 16 inches circumference, within a foot of the middle; if of two marks, 23 inches; of 3, 28 ; of 4, 35 ; of 5, 38 inches about, and so proportionably.

27. Billets were to be of three foot and four inches in length : The single to be 17 inches and an half about ; and every billet of one cast (as they term the mark) to be ten inches about : Of two cast, fourteen inches, and to be marked (unless for the private use of the owner) within six inches of the middle : Of one cast, within four inches of the end, &c.

Every bound faggot should be three foot long; the band twenty four inches circumference, besides the knot.

In the 43 Eliz. the same Statute (which before only concern'd London and its suburbs) was made more universal ; and that of Ed. 6. explain'd with this addition : For such taleshides as were of necessity to be made of cleft-wood, if of one mark and half round, to be 19 inches about ; if quarter-cleft 18 inches $\frac{1}{2}$: Marked two, being round it shall be 23 inches compass ; half-round 27 ; quarter-cleft 26 ; marked three, round 28 ; half-round 33 ; quarter-cleft 32 ; marked four, being round 33 inches about : half-round 39 ; quarter-cleft 38 ; marked five, round 38 inches about ; half-round 44, quarter-cleft 43 ; the measure to be taken within half a foot of the middle of the length mention'd in the former Statute.

Then for the billet, every one nam'd a single, being round, to have 7 inches $\frac{1}{2}$ circumference ; but no single to be made of cleft-wood : If marked one, and round, to contain 11 inches compass ; if half-round 13; quarter-cleft 12 $\frac{1}{2}$.

If marked two, being round, to contain 16 inches; half-round 19; quarter-cleft 18½; the length as in the Statute of King Edward 6.

28. Faggots to be every stick of three foot in length, excepting only one stick of one foot long, to harden and wedge the binding of it : This, to prevent the abuse (too much practis'd) of filling the middle part, and ends with trash and short sticks, which had been omitted in the former Statute : Concerning this and of the dimensions of wood in the stack, see *Copp'ces* cap. 1. book 3. to direct the less instructed purchaser : And I have been the more particular upon this occasion ; because, than our fuel bought in billet by the notch (as they call it in London) there is nothing more deceitful ; for by the vile iniquity of some wretches, marking the billets as they come to the wharf, gentlemen are egregiously cheated. I could produce an instance of a friend of mine (and a member of this Society) for which the wood-monger has little cause to brag ; since he never durst come at him, or challenge his money for the commodity he brought ; because he durst not stand to the measure.

At Hall near Foy, there is a faggot which consists but of one piece of wood, naturally grown in that form, with a band wrapped about it, and parted at the ends into four sticks, one of which is subdivided into two others : It was carefully preserved many years by an Earl of Devonshire, and looked on as portending the fate of his posterity, which is since indeed come into the hands of four Cornish gentlemen, one of whose estates is likewise divided 'twixt two heirs. This we have out of Camden, and I here note, for the extravagancy of the thing ; though as to the verity of such portents from trees, &c. I do not find (upon enquiry, which I have diligently made of my Lord Brereton) that there is any certainty of the rising of those logs in the lake belonging to that place, so as still to premonish the death of the heir of that family, how confidently soever reported; tho' sometimes it has happen'd, but the event is not constant. To this class may be referred what is affirmed concerning the fatal prediction of oaks bearing strange leaves, which may be enquired of : And of accidents fasciating the boughs and branches of trees, Dr. Plot takes notice of in willows and other soft woods, especially in an ash at Bisseter uniformly wreath'd two or three times round : Such a curiosity also hangs up in the portic of the Physic-Garden at Oxford, in a top-branch of holly, which shews it likewise happening sometimes even to harder woods, and 'tis probable that such as we sometimes find so helically twisted, have receiv'd some blast, that has contracted the fibers, and curl'd them in that extravagant manner. Wonderful contorsion and perplexity of the parts of trees, may be seen and admir'd in tea-roots, especially in that given to the Royal Society by the Right Honourable the Lord Summer, (the late most learned President,) amongst the natural rarities of the repositary.

29. But I will now describe to you the mystery of charing, (whereof something was but touch'd in the process of extracting tar out of the pines) as I receiv'd it from a most industrious person, and so conclude the chapter.

There is made of char-coal usually three sorts, viz. one for the iron-works, a second for gun-powder, and a third for London and the Court, besides small-coals, of which we shall also speak in its due place.

We will begin with that sort which is us'd for the iron-works, because the rest are made much after the same manner, and with very little difference.

The best wood for this is good oak, cut into lengths of three foot, as they size it for the stack : This is better than the cord-wood, though of a large measure, and much us'd in Essex.

The wood cut, and set in stacks ready for the coaling, chuse out some level place in the copp'ce, the most free from stubs, &c. to make the hearth on: In the midst of this area drive down a stake for your centre, and with a pole, having a ring fasten'd to one of the extreams (or else with a cord put over the centre) describe a circumference from twenty, or more feet semidiameter, according to the quantity of your wood design'd for coaling, which being near, may conveniently be chared on that hearth; and which at one time may be 12, 16, 20, 24, even to 30 stack: If 12 therefore be the quantity you will coal, a circle whose diameter is 24 foot, will suffice for the hearth; If 20 stack, a diameter of 32 foot ; I 30, 40 foot, and so proportionably.

Having thus marked out the ground, with mattocks, haws, and fit instruments, bare it of the turf, and of all other cumbustible stuff whatsoever, which you are to rake up towards the peripherie, or out-side of the circumference, for an use to be afterwards made of it ; plaining and levelling the ground within the circle : This done, the wood is to be brought from the nearest part where it is stack'd, in wheel-barrows ; and first the smallest of it plac'd at the utmost limit, or very margin of the hearth, where it is to be set long-ways, as it lay in the stack ; the biggest of the wood pitch, or set up on end round about against the small wood,

oo

and all this within the circle, till you come within
five or six foot of the centre; at which distance you
shall begin to set the wood in a triangular form (as
in the following print, *a*) till it come to be three foot
high : Against this again, place your greater wood
almost perpendicular, reducing it from the triangular
to a circular form, till being come within a yard of
the centre, you may pile the wood long-ways, as it
lay in the stack, being careful that the ends of the
wood do not touch the pole, which must now be
erected in the centre, nine foot in height, that so
there may remain a round hole, which is to be form'd
in working up the stack-wood, for a tunnel, and the
more commodious firing of the pit, as they call it, tho'
not very properly.   This provided for, go on to pile,
and set your wood upright to the other, as before ;
till having gain'd a yard more, you lay it long-ways
again, as was shew'd :   And thus continue the work,
still enterchanging the position of the wood, till the
whole area of the hearth and circle be filled and piled
up at the least eight foot high, and so drawn in by
degrees in piling, that it resemble the form of a cop-
ped brown houshold-loaf, filling all inequalities with
the smaller trunchions, till it lie very close, and be
perfectly and evenly shaped.   This done, take straw,
haume, or fern, and lay it on the out-side of the
bottom of the heap, or wood, to keep the next cover
from falling amongst the sticks : Upon this put on
the turf, and cast on the dust and rubbish which was
grubbed and raked up at the making of the hearth,
and reserved near the circle of it ; with this cover the
whole heap of wood to the very top of the pit or
tunnel, to a reasonable and competent thickness, beat-
en close and even, that so the fire may not vent but

in the places where you intend it ; and if in preparing
the hearth, at first, there did not rise sufficient turf
and rubbish for this work, supply it from some convenient place near to your heap : There be who cover
this again with a sandy, or finer mould, which if it
close well, need not be above an inch or two thick :
This done, provide a screene; by making light hurdles
with slit rods, and straw of a competent thickness, to
keep off the wind, and broad, and high enough to
defend an opposite side to the very top of your pit,
being eight or nine foot ; and so as to be easily re-
moved, as need shall require, for the *luing* of your pit.

When now all is in this posture, and the wood
well rang'd, and clos'd, as has been directed, set fire
to your heap : But first you must provide you of a
ladder to ascend the top of your pit: This they usually
make of a curved tiller fit to apply to the convex
shape of the heap, and cut it full of notches for the
more commodious setting the colliers feet, whiles they
govern the fire above ; when now they pull up, and
take away the stake which was erected at the center,
to guide the building of the pile and cavity of the
tunnel. This done, put in a quantity of charcoals
(about a peck) and let them fall to the bottom of the
hearth; upon them cast in coals that are fully kindled;
and when those which were first put in are beginning
to sink, throw in more fuel; and so, from time to
time, till the coals have universally taken fire up to
the top : Then cut an ample and reasonable thick
turf, and clap it over the hole, or mouth of the tunnel, stopping it as close as may be with some of the
former dust and rubbish : Lastly, with the handles of
your rakers, or the like, you must make vent-holes,
or registers (as our chymists would name them)

through the stuff which covers your heap to the very
wood, these in rangers of two or three foot distance,
quite round within a foot (or thereabout) of the top,
tho' some begin them at the bottom : A day after
begin another row of holes a foot and half beneath
the former, and so more, till they arrive to the ground,
as occasion requires.   Note, that as the pit does coal
and sink towards the centre, it is continually to be
fed with short and fitting wood, that no part remain
unfir'd ; and if it chars faster at one part than at
another, there close up the vent-holes, and open them
where need is : A pit will in this manner be burning
off and charing, five or six days, and as it coals, the
smoak from thick and gross clouds, will grow more
blue and livid, and the whole mass sink accordingly;
so as by these indications you may the better know
how to stop and govern your spiracles. Two or
three days it will only require for cooling, which
(the vents being stopped) they assist, by taking now
off the outward covering with a rabil or rubler ; but
this, not for above the space of one yard breadth at a
time ; and first remove the coursest and grossest of it,
throwing the finer over the heap again, that so it may
neither cool too hastily, nor endanger the burning
and reducing all to ashes, should the whole pit be
uncover'd and expos'd to the air at once ; therefore
they open it thus round by degrees.

When now by all the former symptoms you judge
it fully chared, you may begin to draw ; that is, to
take out the coals, first round the bottom, by which
means the coals, rubbish and dust sinking and falling
in together, may choak and extinguish the fire.

Your coals sufficiently cool'd with a very long-
tooth'd rake, and a vann, you may load them into the

coal-wains, which are made close with boards, purposely to carry them to market : Of these coals the
grosser sort are commonly reserv'd for the forges and
iron-works ; the middling and smoother put up in
sacks, and carried by the colliers to London, and the
adjacent towns ; those which are char'd of the roots,
if pick'd out, are accounted best for chymical fires,
and where a lasting and extraordinary blast is requir'd.

30. Coal for the powder-mills is made of alder-
wood (but limetree were much better, had we it in
that plenty as we easily might) cut, stack'd and set
on the hearth like the former : But first, ought the
wood to be wholly disbark'd (which work is to be
done about Midsummer before) and being throughly
dry, it may be coaled in the same method, the heap
or pile only somewhat smaller, by reason that they
seldom coal above five or six stacks at a time, laying
it but two lengths of the wood one above the other,
in form somewhat flatter on the top than what we
have described. Likewise do they fling all their
rubbish and dust on the top, and begin not to cover
at the bottom, as in the former example. In like
sort, when they have drawn up the fire in the tunnel,
and stopp'd it, they begin to draw down their dust
by degrees round the heap ; and this proportionably
as it fires, till they come about to the bottom ; all
which is dispatch'd in the space of two days. One
of these heaps will char threescore sacks of coal,
which may all be carried at one time in a waggon ;
and some make the Court-coals after the same manner.
Lastly,

31. Small-coals are made of the spray and brush-
wood which is shripped off from the branches of
copp'ce-wood, and which is sometimes bound up into

bavins for this use ; though also it be as frequently chared without binding, and then they call it cooming it together : This, they place in some near floor, made level, and freed of incumbrances, where setting one of the bavins, or part of the spray on fire, two men stand ready to throw on bavin upon bavin (as fast as they can take fire, which makes a very great and sudden blaze) till they have burnt all that lies near the place, to the number (it may be) of five or six hundred bavins : But e're they begin to set fire, they fill great tubs or vessels with water, which stand ready by them, and this they dash on with a great dish or scoup, so soon as ever they have thrown on all their bavins, continually plying the great heap of glowing coals, which gives a sudden stop to the fury of the fire, whiles with a great rake they lay, and spread it abroad, and ply their casting of water still on the coals, which are now perpetually turn'd by two men with great shovels, a third throwing on the water : This they continue till no more fire appears, tho' they cease not from being very hot : After this, they shovel them up into great heaps, and when they are throughly cold, put them up in sacks for London, where they use them amongst divers artificers, both to kindle greater fires, and to temper, and aneal their several works : Lastly, this is to be observ'd, that the wood which yields the finest coal, is more flexible and gentle than that which yields the contrary.

32. The best season for the fetching home of other fuel, is from June ; the ways being then most dry and passable, yet I know some good husbands will begin rather in May ; because fallowing, and stirring of ground for corn, comes in the ensuing months, and the days are long enough, and swains have then least to do.

33. And thus we have seen how for house-boot, and ship-boot, plow-boot, hey-boot and fire-boot, the planting and propagation of timber and forest-trees is requisite, so as it was not for nothing, that the very name (which the Greek, generally apply'd to timber) ὕλη, by *synecdoche*, was taken always *pro materia* ; since we hardly find any thing in nature more universally useful ; or, in comparison with it, deserving the name of material ; it being, in truth, as the mother parent and (metaphorically) the passive principle ready for the form.

34. Lastly, to compleat this chapter of the universal use of trees,[1] and the parts of them, something I could be tempted to say concerning staves, wands, &c. their antiquity, use, divine, domestick, civil and politicial ; the time of cutting, manner of seasoning, forming, and other curious particulars (how dry soever the subject may appear) both of delight and profit : but we reserve it for some more fit opportunity, and perhaps, it may merit a peculiar treatise, as acceptable as it will prove divertisant.   Instead of this we will therefore gratifie our reader with some no inconsiderable secrets : And first we will begin with a few plain directions for such persons and country gentlemen, as (being far distant from, or unhandsomely impos'd upon by common painters,) may be desirous to know how to stop, prime and paint their timber-work at home, and save the expence of work by any of their servants indu'd with an ordinary capacity.

Putty to stop the chaps and cracks of wrought timber, is made of white and red-lead, and some Spanish-white (not much) temper'd and bruised with so much lin-seed-oyl as will bring it to the consistence of a past.   Then,

---

[1] See for this Dr. Grew of the vegetation of trunks, cap. 7.

Your first priming shall be of oaker and Spanish-white, very thinly ground : The second with the same, a little whiter ; but it matters not much. The third and last, with white-lead alone ; some mingle a little Spanish-white with it, but it is better omitted. If you desire it exquisite, instead of lin-seed oyl, use that of wallnuts : But the ordinary stone-colour for gross work, expos'd to the air, may be of less expence, with the more ordinary oyl, to which you may add a little charcoal in the grinding : But if (not much minding a small charge) you desire it more fair and durable, lay your work three times with white-lead, (which is indeed much better than Spanish-white) the first and second primer very thin, yet so as not to run : These may be with lin-seed-oyl ; but the last with nut-oyl, and some oyl of terpentine temper'd together, which preserves it from ternishing, and losing colour, (I speak here of work within-doors) : The ordinary priming with red, being a cheat among painters ; seeing white upon white must needs render the colour still whiter and fairer.

If it be for out-work, and expos'd to the air, you may spare the terpentine, whilst nut-oyl through all the three grindings were most desirable.

To vein and wave on white, temper a little lamp-black and white exceedingly thin with nut-oyl and terpentine, and then dipping a gentle flexible feather, vein and undulate your work with a light hand, as naturally as you can, to express the veins of marble, &c. either on black or any other ; but the grain of timber, with a slight of the pensil : Vernish, is often us'd, where they paint in size. For other oyl-colours,

Blew, is made of indigo, with a small addition of red-lead, or verdigriese for a dryer ; unless you will use drying-oyl, which is much preferrable, and is made of lin-seed-oyl boil'd with a little umber bruised small : I speak nothing here of smalt and byce, which is only done by strewing.

Green, with verdigriese ground with lin-seed-oyl pretty thick, and then temper'd with joyners vernish in a glaz'd pot of earth (the best to preserve your colours in) till it run somewhat thin ; and just touch it with your brush, when you lay it on, having prim'd it the second time with white.

There is also a fair grass-green for traillage, priming first with yellow, then with *vert de montagne*, or *lapis Armeniacus.*

Note, that every primer must by dry, before you go it over again.

If you will re-vaile, as they term it, and shadow, or vein your stone-colour, there is a colour call'd shadowing-black ; or you may now and then lightly touch it with a little red-lead ; or work with umber.

It will also behove you to have a good smooth slat, and a pibble mullar well polish'd, which may be bought at London ; as likewise a dozen of large, and lesser brushes, and glaz'd pots; and to grind the colours perfectly well. The Spanish-white requires little labour ; the shadowing black, none at all.

When you have finish'd, wash your brushes with warm-water and a little soap : Preserve your oyl in bladders ; and what colour you leave, plunge the pots into fair-water, so as they may stand a little cover'd in it, which will keep them from growing dry, till you have occasion for them. That you may not be altogether ignorant of the charge and price of the ingredients, which seldom varies :

Clear and sweet lin-seed-oyl is usually had for 4*s.* per gallon.

Spruce-oaker,of all sorts to prime with,3*s*.per pound.

Spanish-white, for half a penny : White-lead 3*d.* per pound.

Vert-de-Greece, clean and bright, 3*s*. per pound.

Black to shadow with, exceeding cheap.

Joiners vernish, 6*d*. per pound.

So as for farther direction ; of white-lead six pound, Span. white six pound, spruce-oaker three pound, vert-de-Greece half a pound, vernish one pound, shadowing-black half a pound, &c. will serve one for a pretty deal of work, and easily inform what quantities you should provide for a greater or lesser occasion.

We will next impart a receipt for a cheap black-dye, such yet as no weather will fetch out, and that may be of use both within and without doors, upon wainscot, or any fine timber, as I once apply'd it to a coach with perfect success.

Take of galls, grosly contus'd in a stone-morter, one pound, boyl them in three quarts of white-wine vinegar to the diminution of one part, two remaining: With this, rub the wood twice over ; then, take of the silk-dyers black, liquid (cheap and easie to be had) a convenient quantity, mix it at discretion with lampblack and *aqua vitae*, sufficient to make it thin enough to pass a strainer : With this, die over your work again ; and if at any time it be stain'd or spotted with dirt, &c. rubbing it only with a wollen cloth dipp'd in oil, it will not only recover, but present you with a very fair and noble polish. There is a black which joyners use to tinge their pear-tree with, and make it resemble ebony, and likewise fir, and other woods for cabinets, picture-frames, &c. which is this :

Take log-wood *q. s.* boil it in ordinary lie, and
with this paint them over : when 'tis dry, work it
over a second time with lampblack and strong size :
That also dry, rub off the dusty sootiness adhering to
it, with a soft brush, or cloth ; then melt some bees-
wax, mixing it with your lamp-black and size, and
when this is cold, make it up into a ball, and rub over
your former black : Lastly, with a polishing-brush
(made of short stiff boars bristles, and fastned with
wyre) labour it till the lustre be to your liking. But,

The black putty, wherewith they stop and fill up
cracks and fissures in ebony, and other fine wood, is
compos'd of a part of the purest rosin, bees-wax and
lamp-black : This they heat and drop into the cran-
nies ; then with an hot iron, glaze it over, and being
cold, scrape it even with a sharp chizel, and after all,
polish it with a brush of bents, a wollen-cloth, felt,
and an hog's-hair rubber : Also mastick alone, mingled with a proper colour, is of no less effect.

35. We conclude all with that incomparable secret
of the *Japon* of China-vernishes, which has hitherto
been reserved so choicely among the *virtuosi* ; with
which I shall suppose to have abundantly gratified
the most curious employers of the finer woods.

Take a pint of spirit of wine exquisitely de-
phlegm'd, four ounces of gum-lacq, which thus cleanse:
Break it first from the sticks and rubbish, and roughly
contusing it in a morter, put it to steep in fountain-
water, ty'd up in a bag of course linnen, together
with a very small morsel of the best Castile-sope, for
12 hours ; then rub out all the tincture from it, to
which add a little alum, and reserve it apart : The
gum-lacq remaining in the bag, with one ounce of
sandrac (some add as much mastic and white-amber)

dissolve in a large matras (well stopp'd) with the
spirit of wine by a two days digestion, frequently
agitating it, that it adhere not to the glass : Then
strain and press it forth into a lesser vessel : Some
after the first infusion upon the ashes, after twenty
four hours, augment the heat, and transfer the matras
to the sand-bath, till the liquor begins to simper ; and
when the upper part of the matras grows a little hot,
and that the gum-lacq is melted, which by that time
(if the operation be heeded) commonly it is, strain it
through a linnen-cloth, and press it 'twixt two sticks
into the glass, to be kept for use, which it will eternal-
ly be, if well stopp'd.

THE APPLICATION.

The wood which you would vernish, should be
very clean, smooth, and without the least freckle or
flaw ; and in case there be any, stop them with a
paste made of gum tragacanth, incorporated with
what colour you design : Then cover it with a layer
of vernish purely, till it be sufficiently drench'd with
it : Then take seven times the quantity of the vernish,
as you do of colour, and bruise it in a small earthen-
dish glaz'd, with a piece of hard wood, till they are
well mingled : Apply this with a very fine and full
pencil ; a quarter of an hour after do it over again,
even to three times successively ; and if every time
it be permitted to dry, before you put on the next,
'twill prove the better : Within two hours after
these four layers (or sooner if you please) polish it
with presle (which our cabinet-makers call, as I think,
Dutch-reeds) wet, or dry ; nor much imports it, tho'
in doing this, you should chance to discover any of

the wood ; since you are to pass it over four or five times, as above ; and if it be not yet smooth enough, presle it again with the reeds, but now very tenderly: Then rub it sufficiently with tripoly, and a little oyl-olive, or water : Lastly, cover it once or twice again with your vernish, and two days after, polish it as before with tripoly, and a piece of hatters felt.

THE COLOURS.

To make it of a fair red, take Spanish vermilion, with a quarter part of Venice lack.

For black, ivory calcin'd (as chymists speak) 'twixt two well luted crucibles, which being ground in water, with the best and greenest copperas, and so let dry, reserve.

For blue, take ultra-marine, and only twice as much vernish as of colour.   The rest are to be applied like the red, except it be the green, which is hard to make fair and vivid, and therefore seldom used.

Note, the right *japon* is done with three or four layers of vernish with the colours ; then two of pure vernish uncolour'd (which is made by the former process, without the sandrac which is only mingled and used for reds) which must be done with a swift and even stroke, that it may not dry before the *aventurin* be sifted on it ; and then you are to cover it with so many layers of pure vernish, as will render it like polish'd glass.   Last of all furbish it with tripoly, oyl, and the felt, as before directed.   Note,

By *venturine* is meant the most delicate and slender golden-wyre, such as embroiderers use, reduc'd to a kind of powder, as small as you can file or clip it : This strewed upon the first layer of pure vernish, when dry, superinduce what colour you please ; and this is prettily imitated with several talkes.

This being the first time that so rare a secret has been imparted (and which since the first publication of it, has been so successfully improv'd amongst our cabinet-makers here in London) the reader will believe that I envy him nothing which may be of use to the publick : And tho' many years since we were master of this curiosity, Athanasius Kircher has set down a process in his late *China Illustrata* pretty faithfully ; yet, besides that it only speaks Latin (such as 'tis) it is nothing so perfect as ours. Howbeit, there we learn, that the most opulent Province of Chekiang is for nothing more celebrated, than the excellent paper which it produces, and the gum call'd ciè (extilling from certain trees) with which they compose their famous vernish, so universally valu'd over the world, because it is found above all other inventions of that nature, to preserve and beautifie wood above any thing which has hitherto been detected : And it has accordingly so generally obtain'd with them, that they have whole rooms and ample chambers wainscotted therewith, and divers of their most precious furniture ; as cabinets, tables, stools, beds, dishes, skreens, staves, frames, pots, and other utensils : But long it was e'er we could for all this, approach it in Europe to any purpose, till F. Eustachius Imart, an Augustine-monk, obtained the secret, and oblig'd us with it.

And now after all, this vernish is said to be improv'd by a later receipt sent from the China missionaries to the Great Duke of Tuscany, and communicated by Dr. Sherards and described in the *Philosophical Transactions*, num. 262. to which I refer the curious both for the materials, colours, composition and working.

I know not whether it may be any service to speak here of coloured woods, I mean such as are naturally so, because besides the berbery for yellow, holly for white, and plum-tree with quick-lime and urine, for red, we have very few : Our inlayers use fustic, locust, or acacia ; Brasile, prince and rose-wood for yellow and reds, with several others brought from both the Indies ; but when they would imitate the natural turning of leaves in their curious comparti- ments and bordures of flower-works, they effect it by dipping the pieces (first cut into shape, and ready to in-lay) so far into hot sand, as they would have the shadow, and the heat of the sand darkens it so gradually, without detriment or burning the thin chip, as one would conceive it to be natural.

Note, that the sand is to be heated in some very thin brasspan, like to the bottom of a scale or bal- lance : This I mention, because the burning with irons, or *aqua-fortis*, is not comparable to it.

I learn also, that soft wood attains little politure without infinite labour, and the expedient is, to plane it often, and every time you do so, to smear it with strong glew, which easily penetrating, hardens it ; and the frequenter you do this, and still plane it, the harder and sleeker it will remain.

And now we have spoken of glew, 'tis so common and cheap, that I need not tell you it is made by boiling the sinews, &c. of sheeps-trotters, parings of raw-hides, &c, to a gelly, and straining it : But the finer and more delicate work is best fastned with fish-glew, to be had of the druggist by the name of *ichthyocolla* ; you may find how the best is made of the skin of sturgeon, in the *Philos. Trans.* vol. 11. num. 129.

36. And here I conclude, summing up all the good qualities, and transcendent perfections of trees, in the harmonious poet's consort of eulogies.

[1] ...... Pines are for masts an useful wood,
Cedar and cypress, to build houses good :
Hence covers for their carts, and spokes for wheels
Swains make, and ships do form their crooked keels ;
With twiggs the sallows, elms with leaves are fraight ;
Myrtles stout spears, and cornel good for fight :
The yews into Ityrean bows are bent ;
Smooth limes, and box, the turners instrument
Shaves into form, and hollow cups does trim ;
And down the rapid Po light alders swim :
In hollow bark bees do their honey stive,
And make the trunk of an old oak their hive.

And the most ingenious Ovid, where he introduces the miraculous groves rais'd by the melodious song of Orpheus,

[2] ...... Nor trees of Chaony,
The poplar, various oaks that pierce the sky,
Soft linden, smooth-rind beech, unmarried bays,
The brittle hasel, ash, whose spears we praise,
Unknotty fir, the solace shading planes,

[1] ......dant utile lignum
Navigiis pinos, domibus cedrumque cupressosque;
Hinc radios trivere rotis, hinc tympana plaustris
Agricolæ, & pandas ratibus posuere carinas.
Viminibus salices, fecundæ frondibus ulmi :
At myrtus validis hastilibus, & bona bello
Cornus : Ityræos taxi torquentur in arcus.
Nec tiliæ leves, aut torno rasile buxum,
Non formam accipiunt ferroque cavantur acuto :
Nec non & torrentem undam levis innatat alnus
Missa Pado, nec non & apes examina condunt
Corticibusque cavis, vitiosæque ilicis alveo :
Georg. 2.

[2] ......non Chaonis abfuit arbor,
Non nemus Heliadum, non frondibus æsculus altis,
Nec tiliæ molles nec fagus, & innuba laurus,
Et coryli fragiles, & fraxinus utilis hastis ;
Enodisque abies, curvataque glandibus ilex,

Rough chesnuts, maple fleck'd with different granes,
Stream-bordering willow, lotus loving lakes,
Tough box, whom never sappy spring forsakes,
The slender tamarisk, with trees that bear
A purple fig, nor myrtles absent were.
The wanton ivy wreath'd in amorous twines,
Vines bearing grapes, and elms supporting vines,
Straight service-trees, trees dropping pitch, fruit-red
Arbutus, these the rest accompanied.
With limber palms, of victory the prize :
And upright pine, whose leaves like bristles rise,
Priz'd by the mother of the gods......
                                        Sandys.

as the incomparable poet goes on, and is imitated by
our divine Spencer, where he brings his gentle knight
into a shady grove, praising

......the trees so straight, and high,
The sailing pine, the cedar proud, and tall,
The vine-prop elm, the poplar never dry,
The builder oak, sole king of forests all ;
The aspine, good for staves ; the cypress funeral :
The laurel, meed of mighty conquerors
And poets sage ; the fir that weepeth still ;
The willow, worn of forlorn paramours :
The yew, obedient to the bender's will ;
The birch for shafts ; the sallow for the mill ;
The myrrh sweet bleeding in the bitter wound ;
The war-like beech ; the ash for nothing ill ;
The fruitful olive ; and the platane round ;
The carver holm ; the maple, seldom inward sound.
                                        Canto 1.

Et platanus genialis, acerque coloribus impar,
Amnicolæque simul salices, & aquatica lotos,
Perpetuóque virens buxus, tenuesque myricæ,
Et bicolor myrtus, & baccis cærula ficus.
Vos quoque flexipedes hederæ venistis, & una
Pampineæ vites, & amictæ vitibus ulmi,
Ornique, & piceæ, pomoque onerata rubenti
Arbutus, & lentæ victoris præmia palmæ,
Et succincta comas, hirsutaque vertice pinus
Grata Deum matri, &c.   *Met.* 10

And in this symphony might the noble Tasso bear likewise his part; but that these are sufficient, & *tria sunt omnia*.

37. For we have already spoken of that modern art of tapping trees in the Spring, by which doubtless some excellent and specific medicines may be attained ; as (before) from the birch for the stone ; from elms and elder against fevers ; so from the vine, the oak, and even the very bramble, &c, besides the wholsom and pleasant drinks, spirits, &c. that may possibly be educed out of them all, which we leave to the industrious, satisfying our selves, that we have been among the first who have hinted and published the ways of performing it.

What now remains, concerns only some general precepts, and directions applicable to most of that we have formerly touched ; together with a brief of what farther laws have been enacted for the improvement and preservation of woods ; and which having dispatched, we shall with a short *paraenesis* touching the present ordering and disposing of the Royal plantations for the future benefit of the nation, put an end to this rustick discourse.

CHAPTER V.

Aphorisms, or certain General Precepts of use to the foregoing Chapters.

1. Try all sorts of seeds, and by their thriving you shall best discern what are the most proper kinds for grounds,

Quippe solo natura subest......

and of these design the main of your plantation. Try all soils, and fit the species to their natures : Beech, hasel, holly, &c. affect gravel and gritty; and if mix'd with loam, oak, ash, elm, &c. In stiff ground the ash, horn-beam, &c. and in a light feeding ground or loam, any sort whatsoever : In the lower and wetter lands, the aquatics, &c.

2. Keep your newly sown seeds continually fresh, and in the shade (as much as may be) till they peep.

3. All curious seeds and plants are diligently to be weeded, till they are strong enough to over-drop or suppress them: And you shall carefully haw, half-dig, and stir up the earth about their roots during the first three years; especially in the vernal and autumnal æquinoxes : This work to be done in a moist season for the first year, to prevent the dust, and the suffocating of the tender buds ; but afterwards, in the more dry weather.

4. Plants, rais'd from seed, shall be thinn'd where they come up too thick ; and none so fit as you thus draw, to be transplanted into hedg-rows, especialy where ground is precious.

Suffragines, nepotes and *traduces* come in here, for general direction ; I begin with

5. Succers, that sprout from the farthest part of the stem, or body of the mother tree, are best, as easier plucked-up without detriment to the roots and fibers, or violence to the mother : It were good therefore first to uncover the roots whence they spring, and to cut them close off, replanting them immediately : Those which grow at more distance, may be separated, with some of the old root, if you find the succher not well furnished.

To produce succers, lay the roots bare, and slit some of them here and there discreetly, and then cover them.

6. Layers, are to be bent down and couched in rich mould, and if you find them stubborn, you may slit a little in the bark and wood, but no deeper than to make it ply, without wounding the tender heart : Putting forth root is assisted by pricking the bark, slitting, or binding a pack-thread about the part you would have them spring from.

The proper season is, from the early spring, or mid-August, &c. and in all dry seasons to keep them diligently watered.

7. Slips, and cuttings (by which most trees may be propagated) taken in moist ground, from August to the end of April frequently moistned ; should be separated at the burs, joints or knobs two or three inches beneath them: Strip them of their leaves before you bury them, leaving no side branches or little top: Some slit the end where it is cut off ; at two years end is the soonest they will be fit to take-up ; layers much sooner.

8. In transplanting, omit not the placing of your trees towards their accustomed aspect : And if you have leisure, make the holes the Autumn before, the wider the better, three foot over, and two deep is little enough if the ground be any thing stiff ; often stirring and turning the mould, and mixing it with better as you may find cause: This done, dig or plough about them, and that as near their stems as you can come, without hurting them, and therefore rather use the spade for the first two or three years; and preserve what you plant steady from the winds and annoyance of cattle, &c.

9. Remove the softest wood to the moistest grounds, as in numb. 1.

Divisae arboribus patriae..........
Georg. 2.

10. Begin to transplant forest-trees when the leaves fall after Michaelmas; you may adventure when they are tarnish'd and grow yellow: It is lost time to commence later, and for the most part of your trees, early transplanters seldom repent; for sometimes a tedious bind of frost prevents the whole season, and the baldness of the tree is a note of deceit; for some oaks, horn-beam, and most beeches, preserve their dead leaves till new ones push them off.

11. Set deeper in the lighter grounds than in the strong; but shallowest in clay: Five inches is sufficient for the dryest, and one or two for the moist, provided you establish them against winds.

12. Plant forth in warm, and moist seasons; the air tranquil and serene; the wind westerly, but never whiles it actually freezes, rains, or in misty weather; for it moulds and infects the roots.

13. What you gather, and draw out of woods, plant immediately, for their roots are very apt to be mortified, or harden'd and wither'd by the winds, and cold air.

14. Trees produc'd from seeds, must have the top-roots abated, (the walnut-tree, and some others excepted, and yet if planted merely for the fruit, some affirm it may be adventur'd on with success) and the bruised parts cut away; but sparing the fibrous, for they are the principal feeders; and those who cleanse them too much, are punished for the mistake.

15. In Spring, rub off some of the collateral buds,

to check the exuberancy of sap in the branches, till
the roots be well establish'd.

16. Transplant no more than you well fence ; for
that neglected, tree-culture comes to nothing : Therefore all young-set trees should be defended from the
winds and sun ; especially the east, and north, till
their roots are fixed ; that is, till you perceive them
shoot ; and the not exactly observing of this article,
is cause of the perishing of the most tender plantations ; for it is the invasion of these two assailants
which does more mischief to our new-set, and less
hardy trees, than the most severe and durable frosts
of a whole Winter.   And here let me add this caution
again; that in planting of trees of stature, for avenues,
or shades, &c. you set them at such distance, as that
they be not in reach of the mansion-house, in case of
being blown down by the winds, for reasons sufficiently obvious : See *History of the Storm*, 26 Nov. 1703.

17. The properest soil, and most natural, apply to
distinct species, *nec vero terrae ferre omnes omnia possunt.*
Yet we find by experience, that most of our forest-
trees grow well enough in the coursest lands, provided
there be a competent depth of mould : For albeit
most of our wild plants covet to run just under the
surface; yet where there is not sufficient depth to cool
them, and entertain the moisture and influences, they
are neither lasting nor prosperous.

18. Wood well planted, will grow in moorish,
boggy, heathy, and the stoniest grounds : Only the
white, and blew clay (which is commonly the best
pasture) is the worst for wood, and such good timber
as we find in any of these (oaks excepted) is of an
excessive age, requiring thrice the time to arrive at
their stature.

19. If the season require it, all new plantations are to be plied with waterings, which is better pour'd into a circle at some distance from the roots, which should continually be bared of grass, and if the water be rich, or impregnated, the shoots will soon discover it ; for the liquor being percolated through a quantity of earth, will carry the nitrous virtue of the soil with it ; by no means therefore water at the stem; because it washes the mould from the root, comes too crude, and endangers their rotting : But,

20. For the cooling and refreshing tree-roots, the congesting of rotten litter sprinkl'd over with fine earth is good, or place potsheards, flints, or pibbles near the foot of the stem, for so the poet :

[1] Lime-stones, or squalid shells, that may the rain,
Vapours, and gliding moisture entertain.

But remember you remove them after a competent time, else the vermine, snails, and insects which they produce and shelter, will gnaw, and greatly injure their bark, and therefore to lay a coat of moist rotten litter with a little earth upon it, will preserve it moist in Summer, and warm in Winter, inriching the showers and dews that strain through it.

21. Young plants will be strangled with corn, oats, pease, or hemp, or any rankly growing grain, if a competent circle, and distance be not left (as of near a yard, or so) of the stem ; this is a useful remark : But whether the setting, or sowing of beanes near trees, make them thrive the more (as Theophrastus

[1] Aut lapidem bibulum, aut squallentes infode conchas,
Inter enim labentur aquae, tenuisque subibit
Halitus.........
                                        Georg. 2.

writes, I suppose he means fruit-trees) I leave to experience.   Pythagoras we know prohibited the eating
of them to women.

22. Cut no trees (especially having an eminent
pith in them, being young and tender too) when
either heat or cold are in extreams ; nor in very wet
or snowy weather ; and in this work it is profitable
to discharge all trees of unthriving, broken, wind-
shaken browse, and such as our law terms *cablicia*,
and to take them off to the quick,

................ne pars sincera trahatur.

And for evergreens, especially such as are tender,
prune them not after planting, till they do *radicate*,
that is, by some little fresh shoot, discover that they
have taken.

23. Cut not off the top of the leading-twig or shoot
(unless very crooked, and then at the next erect bud)
when you transplant timber-trees, but those of the
collateral you may shorten, stripping up the rest close
to the stem: and such as you do spare, let them not
be the most opposite, but rather one above another
to preserve the part from swelling, and hindring its
taper growth : Be careful also to keep your trees from
being over top-heavy, by shortning the side branches
competently near the stem : Young plants nipt either
by the frost or teeth of cattle do commonly break on
the sides, which impedes both growth and spiring :
In this case, prune off some, and quicken the lead-
ing-shoot with your knife, at some distance beneath
its infirmity : But if it be in a very unlikely condition
at Spring, cut off all close to the very ground, and
hope for a new shoot, continually suppressing what-

ever else may accompany it, by cutting them away in Summer.

24. Walnut, ash, and pithy-trees are safer prun'd in Summer and warm weather, than in the spring, whatever the vulgar fancy.   And so

I will conclude with the tecnical names, or dis-similar parts of trees, as I find them enumerated by the industrious and learned Dr. Merett. *Scapus, truncus, cortex, liber, malicorium, matrix, medulla & cor, pecten, circuli, surculi, rami, sarmenta, ramusculi, spadix, vimen, virgultum & cremium, vitilia, talea, scobs, termes, turiones, frondes, cachryas & nucamentum, julus & catulus, comae :* The species *frutex, suffrutex, &c.* to which add, *alburnum, capitulum, cima, echinus, geniculum, locustae, pericarpum, petiolus, sugilta, &c.* all which I leave to be put into good and proper English, (as our learned phytologist Mr. Ray has done) by those who shall once oblige our nation with a full and absolutely compleat dictionary, as yet a desiderate amongst us, however of late infinitely improv'd.

To this I shall add, the time and season of the flourishing of trees, computing from the entry of each month as the figures denote ; that is, from March (where the Doctor begins) inclusively. March, *acer,* 3. (i. e.) from March to May, viz. one month ; *& sic de ceteris) populus* 2. *quercus* 5. *sorbus* 2. *ulmus* 2. April, *alnus* 2. *betula* 2. *castanea* 4. *euonymus* 2. *fagus* 2. *fraxinus* 2. *nux-juglans* 3. *salix* 2. *sambucus* 2. May, *cornus* 2. *genista* 4. *juniperus, morus* 2. *tilia* 4. June, *aquifolium* 2. July, *arbutus* 2. Feb. *buxus* 2, *&c.*

Many more useful observations are to be collected, and added to these, from the diligent experience of planters.

CHAPTER VI.

Of the Laws and Statutes for the Preservation, and Improvement of Woods and Forests.

1. 'Tis not to be passed by, that the very first law we find which was ever promulg'd, was concerning trees ; and that laws themselves were first [1] written upon them, or tables compos'd of them ; and after that establishment in Paradise, the next we meet withal are as ancient as Moses ; you may find the statute at large in Deut. c. 20. v. 19, 20. Which though they chiefly tended to fruit-trees, even in an enemies country, yet you will find a case of necessity, only alledg'd for the permission to destroy any other.

2. To sum up briefly the laws, and civil constitutions of great antiquity, by which Servius informs us 'twas no less than capital, *alienas arbores incidere ;* the Lex Aquilia, and those of the xii. Tabb. mention'd by Paulus, Cujas, Julianus, and others of that robe, repeated divers more.

It was by those sacred constitutions provided, that none might so much as plant trees on the confines of his neighbour's ground, but he was to leave a space of at the least five foot, for the smallest tree, that they might not injure him with their shadow. *Si arbor in vicini agrum impenderit, eam sublucato, &c.* and if for all this, any hung over farther, 'twas to be stripp'd up fifteen foot : And this law Balduinus, Olderdorpius,

---

[1] The laws of Numa first cut in *quernis tabulis*, before they were engraven in brass : See Dionysius Halicarnass., lib. 3.

and Hotoman recite out of Ulpian l. 1. *f. de arb.
caedend.* where we have the Prætors interdict express'd,
and the impendent wood adjudged to appertain to
him whose field or fence was thereby damnified :
Nay, the wise Solon prescribed ordinances for the
very distances of trees ; as the divine Plato did
against stealing of fruit, and violating of plantations :
And the interdiction *de glande legenda* runs thus in
Ulpian, *ait prætor, glandem, quæ ex illius agro in tuum
cadit, quo minus illi tertio quoque die legere auferre liceat,
vim fieri veto.* And yet, though by the Praetors permission he might come every third day to gather
it up without trespass, his neighbour was to share of
the mast which so fell into his ground; and this chapter
is well supplied by Pliny, l. 16. c. 5. and Cujas upon
the place, interprets *glandem* to signifie not the acorn
of the oak alone, but all sorts of fruit whatsoever,
l. 136. *f. de verb. signif. l. usus ff. de glande leg.* as by
usage of the Greeks, amongst whom ἀκρόδρυα imports
all kind of trees.

There were also laws concerning boundaries, to be
found at large in other learned authors, *De re agraria*,
of which we give this short extract: Some admitting
any sort of trees, others of peculiar kinds, for the
fencing of their grounds ; others with foreign trees,
that the difference of the wood might serve as a
mark : Some by agreement planted them in common
upon the very borders ; some at their private charge,
a little within the margins of their own fields, &c.
Amongst the different sorts of trees, we find pines
and cypress-trees plac'd for bounds, in others ash,
elm, or poplar ; which being near the limits, with
any cultivated ground between, the intermediate
spaces were fill'd with shrubs. In case the trees were

in common, some preserv'd them untouch'd on both
sides; others, the stems only, lop, tops, and branches,
(especially if they belonged to a particular person) to
cut or spare at their pleasure, provided they planted
others in their room.　In trees marked, it must be
consider'd whether they are in common, which ought
to be marked in the middle, or on each side; and if
one side of the tree have leaves, the other should be
cut, to signify their belonging to those persons, on
the border of whose grounds they are left intire.　To
this for trees 8 foot asunder: Those at 20 foot distance
were marked with X, or Γ, to notify a flexure or
turning there-about: Some permit them to stand till
they arrive to such a bulk and stature as to over-top
the rest, distinguish'd also from those marked on both
sides, whether they stand in woods, barren, or uncultivated land, as being suppos'd in common.　The
same rule holds if marked in the middle : If but one
side be marked, the unmarked side is the boundary :
If the mark be different on either side, (and none else
to be seen) such trees are not to be accounted boundaries : If as sometimes briars and such shrubs grow
on the ancient limits, it must be consider'd of what
kind they are, and should be enquired how it happens
that they are often found in the middle of the fields.
Lastly, in *campagne* and open places, foreign trees
were usually planted.　There are more of those nice
rules to be found among the lawyers, whilst before
any of these instances, the images of Satyrs bounded
the confines, and were counted as *termini*, which none
might remove, without being accounted as sacrilegious, and the person punished with death.　These,
and the *Hermae* were reputed protectors of such
boundaries..........

..............Et te pater
Silvane, tutor finium.   Hor.

In the mean time, no trees whatsoever might be planted near publick aquaducts, lest the roots should insinuate into, and displace the stones : Nor on the very margent of navigable rivers, lest the boats and other vessels passing to and fro, should be hindred, and therefore such impediments were call'd *retae*, *quia naves retinent*, says the *Gloss ;* and because the falling of the leaves corrupted the water. So nor within such a distance of high-ways (which also our own laws prohibit) that they might dry the better, and less cumber the traveller. Trees that obstructed the foundation of houses were to be fell'd ; Barthol. l. 1. *doct. c. de interdict.* Ulp. *in l. priore ff. de arborum caedend.* Trees spreading their roots in neighbour-ground, to be in common ; see Cujas and Paulus in *L. Arb. ff. de communi dividend.*, where more of the alienation of trees fell'd, and not standing but with the funds, as also of the usu-fruit of trees, and the difference 'twixt *arbores grandes*, and *cremiales* or *caeduae*, of all which Ulpian, Baldus, Alciat, with the laws to govern the *conlucatores* and *sublucatores*, and pruners ; *vide Pan. s. c. Sent.* l. 5. Festus, &c. for we pass over what concerns vines and olive-trees, to be found in Cato *de R. R. &c.* Nor is it here that we design to enlarge, as those who have philologiz'd on this occasion *de sycophantis*, and other curious criticisms ; but to pass now on, and confine my self to the prudent sanctions of our own Parliaments: For though according to the old and best spirit of true English, we ought to be more powerfully led by Royal example, than to have need of more cogent and violent laws ;

yet that our discourse may be as ample, and as little defective as we can render it, something 'tis fit should be spoken concerning such laws and ordinances as have been from time to time constituted amongst us for the encouragement and direction of such as do well, and for the animadversion and punishment of those who continue refractory.

But before we descend to our municiple, and present laws and constitutions, let us enquire what was anciently meant by a forest. (Waving those, I think, impertinent etymologies, *quia foris est*, *(Lumbard Gloss, &c.)* a forest is properly an harbour for wild beasts : *quasi ferarum statio ;* for which, mighty tracts and portions of land have been afforded (as the term is) by the kings and monarchs of this nation, beyond any other in Europe, and guarded with such strict, rigorous, and severe laws, as did not extend to the prohibition of killing and destruction of deer and venison alone ; but even to that of killing little silly birds ; and that not only to the forfeiture and loss of goods, but of limb and life. Such, among others, was that of Richard the First, upon incurring the loss of the offender's eyes and testicles, &c. to the unsufferable hindrance of great improvements ; whilst there might have been not only enough for royal diversion, but for the increase of timber and people; which are the true glory and safety of this nation. In the mean time, 'tis remarkable that William Rufus (successor to the great Conqueror) chasing a stag under a spreading oak, was by the glance of an arrow levell'd at the beast, depriv'd of his life. The historian recounts it as God's visiting the sin of the father upon the children, for his demolishing so many churches and villages, and turning them into receptacles and dens

of wild beasts ; there having besides this prince been
two more who met with their death in New-Forest.
There were in Yorkshire alone, in the time of Henry
the VIII[th.] two hundred seventy and five woods (besides
the parks and chases) most of them containing five
hundred acres: See Mr. Camden's *Brit.* As to what we
call wood-land, I know not how to distinguish forest
from woods, unless for its being applicable to all sorts
in common ; for heretofore (which as Strabo tells us)
the ancient inhabitants of this island's security, was
their woods instead of cities and towns, as still they
are among the people of the uncultivated America :
Nor doubtless was our superb, and stately metropolis
(the ancient Trinovant) any other ; from whence
some derive its name, turning *den* only into *don* ;
whilst since our own remembrance, the whole city
was (till the late dreadful conflagration) a wooden
city, almost entirely built of wood and timber.

Wood-land in Warwickshire (says the same learned
antiquary) was anciently call'd *Ardena*, importing
the same in British, and still retaining the same, in
what is left of that vast forest, the Ardenner-Wald
in the Nether Germany, which stretching thro' the
Caledonium of Luxemburg to the confines of Cham-
pagn, for more than an hundred miles in length, was
no more than such as might compass a wood-land ;
from whence our own Danica Silva (the Forest of
Deane) might probably derive its name contracted,
and *Diana Nemorensis* found under the British Ar-
duena and Arden : But dismissing these conjectures,
we now come to the subject of this Chapter, as it
more immediately concerns our Common Law, (and
some of other nations) which we shall deduce in this
order.

3. From the time of Edward the Fourth, were enacted many excellent laws for the planting, securing, cutting, and ordering of woods, copp'ces, and under-woods, as then they took cognizance of them; together with the several penalties upon the infringers; especially from the 25. of Hen. 8. 17, &c. confirm'd by the 13 and 27. of Q. Eliz. *cap.* 25, 19, &c. which are diligently to be consulted, revived, put in execution, and enlarg'd where any defect is apparent; as in particular the Act of exempting of timber of 22 years growth from tithe, for a longer period, to render it compleat, and more effectual to their improvement: And that law repealed, by which willows, sallows, oziers, &c. which they term *Sub-bois*, are reputed but as weeds.

4. Severer punishments have lately been ordain'd against our wood-stealers, destroyers of young trees, &c. By an ancient law of some nation, I read he forfeited his hand, who beheaded a tree without permission of the owner; and I cannot say they are sharp ones, when I compare the severity of our laws against mare-stealers; nor am I by inclination the least cruel; but I do affirm, we might as well live without mares, as without masts and ships, which are our wooden, but no less profitable horses.

5. And here we cannot but perstringe those riotous assemblies of idle people, who under pretence of going a Maying, (as they term it) do oftentimes cut down and carry away fine straight trees, to set up before some ale-house, or revelling place, where they keep their drunken *Bacchanalia*: For though this custom was, I read, introduc'd by the Emperor Anastasius, to abolish the *gentile majana* of the Romans at Ostia; which was to transfer a great oaken-tree out

of some forest into the town, and erect it before their
mistresses door ; yet I think it were better to be
quite abolish'd amongst us, for many reasons, besides
that of occasioning so much wast and spoil as we
find is done to trees at that season, under this wanton
pretence, by breaking, mangling, and tearing down
of branches, and intire arms of trees, to adorn their
wooden idol. The imperial law against such disorders
we have in l. ob. id. ff. *ad legem Aquill.* & in ff. l. 43.
tit. 7. *Arborum furtim caesarum :* See also Triphon. l.
ig. *de bon. off. cont. tab. vel in ligna focaria,* L. Ligni.
ff. *de lege* 3, &c.

To these I might add the laws of our king Ina ; or
as the learned Lambard reckons them in his Ἀρχαιονομία
de priscis Anglorum legibus, whose title is, Be thuthu
baprete : Of burning trees : The sanction runs thus :

If any one set fire of a fell'd wood, he shall be
punished, and besides pay three pounds, and for those
who clandestinely cut wood (of which the very sound
of the ax shall be sufficient conviction) for every tree
he shall be mulcted thirty shillings. A tree so fell'd,
under whose shadow thirty hogs can stand, shall be
mulcted at three pounds, &c.

6. I have heard, that in the great expedition of
88, it was expresly enjoin'd the Spanish commanders
of that signal Armada ; that if when landed they
should not be able to subdue our nation, and make
good their conquest ; they should yet be sure not to
leave a tree standing in the Forest of Dean : It was
like the policy of the Philistines, when the poor
Israelites went down to their enemies smiths to

---

[1] Severe laws against woodstealers, v. Greeneway, *de ll. abrog. in Hollandia*
ad tit. *arbor. furt. caesar.* L. 2. One cruelly whipt for it. See also Carpzovius in
Prax. Crim. par. 2. Quest. 83. Num. 2. seqq. and several others to that purpose.

146 SYLVA BOOK III

sharpen every man his tools; for as they said, lest the
Hebrews make them swords, or spears ; so these, lest
the English build them ships, and men of war :
Whether this were so, or not, certain it is, we cannot
be too jealous for the preservation of our woods; and
especially of those eminent, and with care, inex-
haustible magazines. In the Duke of Luxemburg's
country, no farmer is permitted to fell a timber-tree
without making it appear he hath planted another.
And we have already mention'd that inviolable custom
about Frankford, where the young farmer must
produce a certificate of his having set a number of
walnut-trees, before he have leave to marry : But of
these, and the like, v. Follar in *Constit. Rey. de Offic.
Tract.* 11, 92, 93, &c. I dare not suggest the encouragement of a yet farther restraint, that even
proprietors themselves should not presume to make
havock of some of their own woods, to feed their
prodigality, and heap fuel to their vices ; but it is
worthy of our observation, that (in that inimitable
oration, the second *Philippic*) Cicero does not so
sharply reproach his great antagonist for any other
of his extravagancies (which yet he there enumerates)
as for his wasteful disposure of certain wood-lands
belonging to the Commonwealth, amongst his jovial
bravo's, and lewd companions ; *tua ista detrimenta sunt*
(meaning his debauchees) *illa nostra* ; speaking of the
timber : And doubtless, the spoil and wasting of this
necessary material is no less than a publick calamity ;
this, John Duke of Lancaster knew well enough, when
to revenge the depradations made upon the English
borders, 'tis said, he set four and twenty thousand
axes at work at once, to destroy the woods in Scotland.

7. But to the laws : It were to be wish'd that our

tender and improvable woods, should not admit of
cattle by any means, till they were quite grown out
of reach ; the statutes which connive at it, in favour
of custom, and for the satisfying of a few clamorous
and rude commoners, being too indulgent ; since it is
very evident, that less than a 14 or 15 years enclosure,
is in most places too soon ; and our most material
trees would be of infinite more worth and improvement, were the standards suffer'd to grow to timber,
and not so frequently cut, at the next felling of the
wood, as the general custom is.   In 22 Edw. 4. the
liberty arriv'd but to seven years after a felling of a
forest or purlieu ; and but three years before, without
special licence : This was very narrow ; but let us
then look on England as an over-grown country.

8. Wood in parks was afterwards to be four years
fenced, upon felling ; and yearling colts, and calves
might be put into inclosed woods after two : By the
13 Eliz. five years, and no other cattle till six, if the
growth was under fourteen years ; or until eight, if
exceeding that age till the last felling : All which
statutes being by the Act of Hen. 8. but temporal,
this Parliament of Elizabeth thought fit to make
perpetual.

9. Then, to prevent the destructive razing and
converting of woods to pasture : No wood of two
acres, and above two furlongs from the Mansion-
House, should be indulg'd : And the prohibitions are
good against assarts made in forests, &c. without
licence : The penalties are indeed great ; but how
seldom inflicted ?   And what is now more easie, than
compounding for such a licence ?

In some parts of Germany, where a single tree is
observ'd to be extraordinary fertile, a constant and

plentiful mast-bearer; there are laws to prohibit their felling without special leave: And it was well enacted amongst us, that even the owners of woods within chases, should not cut down the timber without view of officers; this Act being in affirmance of the Common-law, and not to be violated without prescription : See the case cited by my Lord Cook in his *Comment* on Littleton. *Tenure Burgage*. l. 2. Sect. 170. Or if not within chases, yet where a common-person had liberty of chase, &c. and this would be of much benefit, had the regarders perform'd their duty, as 'tis at large described in the writ of the 12 Articles; and that the surcharge of the forests had been honestly inspected with the due perambulations, and ancient metes: Thus should the justices of Eire dispose of no woods without express commission, and in convenient places: *Minuti blaterones quercuum, culi, & curbi*, as our law terms wind-falls, dotterels, scrags, &c. and no others.

10. Care is likewise by our laws to be taken that no unnecessary imbezlement be made by pretences of repair of paling, lodges, browse for deer, &c. wind-falls ; root-falls ; dead and sear-trees, all which is subject to the inspection of the warders, justices, itinerants, &c. and even trespasses done *de viridi* on boughs of trees, thickets, and the like; which (as has been shew'd) are very great impediments to their growth and prosperity, and should be duly looked after, and punished ; and the great neglect of Swain-mote-Courts reformed, &c. see *Consuet. & Assis. fores. pannagium*, or *Pastura pecorum & de glandibus, fleta*, &c. Manwood's *Forest-laws* : Cook *pla. fol.* 366. li. 8. fol. 138.

11. Finally, that the exorbitance and increase of devouring iron-mills were looked into, as to their

distance and number near the seas, or navigable rivers; and what if some of them were even remov'd into another world ? the Holy-Land of New-England, (there to build ships, erect saw-mills, near their noble rivers) for they will else ruin Old-England : Twere better to purchase all our iron out of America, than thus to exhaust our woods at home, although (I doubt not) they might be so order'd, as to be rather a means of conserving them. There was a statute made by Queen Eliz. to prohibit the converting of timber-trees to coal, or other fuel for the use of iron-mills ; if the tree were of one foot square, and growing within fourteen miles of the sea, or the greater rivers, &c. 'Tis pity some of those places in Kent, Sussex and Surrey were excepted in the proviso, for the reason express'd in a statute made 23 Eliz. by which even the employing of any under-wood, as well as great trees, was prohibited within 22 miles of London, and many other navigable rivers, creeks and other lesser distances from some parts of Sussex-downs, Cinque-ports, havens, &c.

One Simon Sturtivant had a patent fromK . James I. 1612. pretending to save 300000 l. a year, by melting iron ore, and other metals, with pit-coal, sea-coal, and brush-fuel ; 'tis pity it did not succeed.

There are several acres of wood-land, of no mean circuit near Rochester, in the county of Kent, extending as far as Bexley, and indeed, for many miles about Shooter's Hill, near the river of Thames, which, were his Majesty owner of, might in few years be of an unvaluable improvement and benefit, considering how apt they are to grow forest, and how opportune they lie for the use of the Royal Navy at Chatham.

12. But yet to prove what it is to manage woods

discreetly ; I read of one Mr Christopher Darell a
Surrey gentleman of Nudigate, that had a particular
indulgence for the cutting of his woods at pleasure,
though a great iron-master; because he so ordered
his works, that they were a means of preserving even
his woods; notwithstanding those unsatiable devourers:
This may appear a paradox, but it is to be made out ;
and I have heard my own father (whose estate was
none of the least wooded in England) affirm, that a
forge, and some other mills, to which he furnished
much fuel, were a means of maintaining and improving his woods; I suppose, by increasing the industry
of planting, and care ; as what he left standing of
his own planting, enclosing and cherishing, lately in
the possession of my most honoured brother George
Evelin of Wotton in the same county, (and now
in mine) did (before the late hurricane) sufficiently
evince; a most laudable monument of his industry,
and rare example, for without such an example, and
such an application, I am no advocate for iron-works,
but a declared denouncer: But nature has thought fit
to produce this wasting oar more plentifully in wood-
land, than any other ground, and to enrich our forests
to their own destruction,

> [1] O poverty, still safe ! and therefore found
> Insep'rably with mischiefs under ground !
> Woods tall, and reverend from all time appear
> Inviolable, where no mine is near.

for so our sweet poet deplores the fate of the Forest
of Dean.

[1] O semper bona pauperies ! & conditus altâ
Thesaurus tellure nocens ! O semper ovantes,
Integræ. salvæque solo non divite Silvæ !
Couleii Pl. l. 6.

13. The same act we have confirmed and enlarged in the twenty seventh of Queen Eliz. Cap. 19. for the preserving of timber-trees, and the penalties of impairing woods much increased ; the tops and offal only permitted to be made use of for this employment.

Nay, our own law makes it wast to cut down high-trees (tho they be not properly timber) standing for safe-guard and defence of a mansion-house) tho it be done for necessary repairs; whilst yet many (and with reason) hold it un-healthful to suffer a dwelling to be choak'd with trees, for want of free passage to the air : To remedy this, there needs only a competent distance to be left void.   But, as a noble [1] person observes, people in these days are so dispos'd to quarrel with timber, as there shall need no advice to demolish trees about their houses upon this account ; In the mean time, as to the incroachment of trees so near our dwellings, for the freer intercourse of air, the late dreadful *silvifragi* storms have cleans'd those places by a remedy worse than the disease, sufficient to deter us from planting not only too near our habitations, but from priding our selves in our more stately avenues, the late boasts of our seats, as by sad experience my self and thousands more have found, that there is nothing stable in this world, which invisible spirits cannot subvert and demolish, when God permits them to do mischief, and convince those who believe there are none, because they do not see, though they feel their effects.

14. As to the law of tithes, I find timber-trees pay none, but others do, both for body, branches, bark, fruit, root, [2] and even the suckers growing out of

[1] Lord North, *Oeconom.*
[2] See L. Bp. of Worcester concerning tithes of parochial clergy, p. 268.

them ; and the tenth of the body sold, or kept : And so of willows, sallows and all other trees not apt for timber : Also of *silva caedua*, as copp'ces, and under-woods, pay the tenth whenever the proprietor receives his nine parts. But if any of these we have named un-exempted are cut only for mounds, fencing, or plow-boot within the parish in which they grow, or for the fuel of the owner, no tithes are due, though the vicar have the tithe-wood, and the parson that of the places so enclosed ; nor are under-woods grubb'd up by the roots tithable, unless for this, and any of the former cases there be prescription. But for timber-trees, such as oak, ash, elm (which are accounted timber in all places after the first twenty years) also beech, horn-beam, maple, aspen, and even hasel (many of which are in some countries reputed timber) they are not to pay tithes, unless they are fell'd before the said age of twenty years from their first planting.

Some think, and pretend, that no tithe is due where is no annual increase, as corn and other grain, hay, and fruit of trees, and some animals ; and that there-fore *silva caedua*, (till it become timber) is exempted : But a Parliament at Sarum did make it titheable, in which are named, even willows, alder, beech, maple, hasel, &c.

In the Wild of Sussex, tithe-wood is not paid, as for faggots ; but in the Downs they pay for both, as I am told ; at which I wonder, there being so little wood at all upon them, or likely to have ever been. Note here,

If the owner fell a fruit-tree (of which the parson has had tithe that year) and convert the wood into fuel, the tithe shall cease ; because he cannot receive the tithe of one thing twice in one year.

Beech, in countries where it abounds, is not tithable; because in such places 'tis not accounted timber.   16 Jac. Co. B., Pinder's Case.

Cherry-trees in Buckinghamshire have been adjudged timber, and tithe-free. Pasch. 17 Jac. B.R.

If a tree be lopp'd under twenty years growth, and afterwards be permitted to grow past twenty years, and then be lopp'd again, no tithe is due for it, tho at the first cutting it were not so.

If wood be cut for hedges, which is not tithable, and any be left of it unemploy'd, no tythe shall be paid for it.

If wood be cut for hop-poles (where the parson or vicar has tithe-hops) in this case he shall not have tythe of hop-poles.

If a great wood consist chiefly of under-wood tithable, and some great trees of beech, or the like grow dispersedly amongst them ; tithe is due, unless the custom be otherwise, of all both great and lesser together : And in like manner, if a wood consist for the most part of timber-trees, with some small scatterings of underwood amongst them, no tithe shall be paid for the under-wood or bushes.   Frin. 19 Jac. B. R. Adjudg. 16 Jac. in C. B., Leonard's Case.

No tithe is to be paid of common of estovers, or the wood burnt in ones house.   Now as to the manner of payment :

To give the parson the tenth acre of wood in a copp'ce, or the tenth cord (provided they are equal) is a good payment, and setting forth of tithe, especially if the custom confirm it.

The tithe of mast of oak, or beech, if sold, must be answer'd by the tenth penny : If eaten by swine, the worth of it.   And thus much we thought fit to

add concerning predial tithes ; who has desire to be
farther informed may consult *Carta de Foresta,* with
Manwood's *Treatise of Forest-Laws* : Cromate on my
Lord Cook's *Rep.* 11. 48, 49, 81. Plow. 470. Brown-
low's *Rep.* 1 part 94. 2 part 150. D. and St. 169,
&c. and that very useful, as well as compendious
English Historical Library, Part III, chap. 4. lately
published by the worthy arch-Deacon, now bishop
of Carlisle.   But let us see what others do.

15. The King of Spain has near Bilboa, sixteen
times as many acres of copp'ce-wood as are fit to be
cut for coal in one year ; so that when 'tis ready to
be fell'd, an officer first marks such as are like to
prove ship-timber, which are let stand, as so many
sacred and dedicate trees ; by which means the iron-
works are plentifully supplied in the same place,
without at all diminishing the stock of timber.   Then
in Biscay again, every proprietor plants three for one
which he cuts down ; and the law obliging them is
most severely executed ; see what we have already
mentioned of the Duke of Luxemburg in this chapter,
and that of the walnut-tree.   There indeed are few,
or no copp'ces ; but all are pollards ; and the very
lopping (I am assur'd) does furnish the iron-works
with sufficient to support them.

16. What the practice is for the maintaining of
these kind of plantations in Germany and France,
has already been observed to this illustrious Society
by the learned Dr. Merret ; *viz* that the Lords and
(for the Crown-lands) the King's Commissioners,
divide the woods, and forests, into eighty partitions ;
every year felling one of the divisions ; so as no
wood is felled in less than fourscore years : And when
any one partition is to be cut down, the officer, or

Lord contracts with the buyer, that he shall at the distance of every twenty foot (which is somewhat near) leave a good, fair, sound and fruitful oak standing. Those of 'twixt forty and fifty years they reckon for the best, and then they are to fence these trees from all sorts of beasts, and injuries, for a competent time ; which being done at the season, down fall the acorns, which (with the autumnal rains beaten into the earth) take root, and in a short time furnish all the wood again, where they let them grow for four or five years, and then grub up some of them for fuel, or transplantations, and leave the most probable of them to continue for timber.

17. The French King permits none of his oak woods, tho belonging (some of them) to *Monsieur* (his royal brother) in appenage, to be cut down ; till his own surveyers and officers have first marked them out; nor are any fell'd beyond such a circuit : Then are they sufficiently fenc'd by him who buys ; and no cattle whatsoever suffered to be put in, till the very seedlings (which spring up of the acorns) are perfectly out of danger.   But these, and many other wholsome ordinances, especially, as they concern the Forest of Dean, we have comprized in the late Statute of the twentieth of his Majesty's reign, which I find enacted five years after the first edition of this *Treatise :* And these laws are worthy our perusal ; as also the Statute prescribing a scheme of proportions for the several scantlings of building timber (besides what we have already touched, Chap. IV. Book III. &c.) which you have 19 Car. II. entituled, An Act for the re-building of London ; to which I refer the reader.

In the mean time, commissioners made purveyers for timber (tho for the King's use) cannot by that authority take timber-trees growing upon any man's free-hold, it being prohibited by Magna Charta : Cap. 21. *Nos nec Ballivi nostri, nec alii, cupimus boscum alienum ad castra, vel ad alia agenda nostra, nisi per voluntatem cujus boscus ille fuerit.*

We might here enlarge this title, by shewing how different the forest-laws are from the common-laws of England, both as to their antiquity and extream severity against all offenders, (of what degree soever) till the oppression was somewhat qualified by the *Charta de Foresta*, and afterwards by yet more favourable [1] concessions; since indeed, our Kings, after the rigor and example of the stern northern princes, rendred it intolerable : But because much of this concerned the preserving Royal game; when as to timber-trees (like Germany) the whole island was almost but one vast forest, and wood so abounding, that what people might have had almost for carrying off the ground it grew on, is now grown so scarce, in those very places, as that fuel is sold by weight : I think Mr. Camden mentions Oxfordshire; even so long since : And here I might mention that vast Caledonian forest, heretofore in Scotland (whence the sea has its name), and the people Caledonians, having now not so much as a single tree to shew for it. Have we not then the greatest reason in the world to take all imaginable care for the preservation and improvement of this precious material ?

We have said nothing of the laws against wood-stealers, (especially those who cut up to the very roots, the most hopeful and thriving oaks, and sell

---

[1] *Assises Forestæ, &c.*

bundles of them for walking-staves, &c.) severely [1] punished in other countries, but leave the rest to our learned in the laws, craving pardon for the errors I may have fallen into, by presuming to discourse of matters out of my element and profession.

CHAPTER VII.

The paraenesis and conclusion, containing some encouragements and proposals for the planting and improvement of his Majesty's forests, and other amœnities for shade, and ornament.

1. Since our forests are undoubtedly the greatest magazines of the wealth and glory of this nation; and our oaks the truest oracles of its perpetuity and happiness, as being the only support of that navigation which makes us fear'd abroad, and flourish at home: It has been strangely wonder'd at by some good patriots, how it comes to pass that many gentlemen have frequently repaired, or gained a sudden fortune, with plowing part of their parks, and letting out their fat grounds to gardeners, &c. and very wild wood-land parcels (as may be instanced in several places) to dressers of hop-yards, &c. whiles the royal portion lies folded up in a napkin, uncultivated, and neglected: especially those great and ample forests; where, tho plowing and sowing have been forbidden, a Royal command and design may well dispense with it, and

---

[1] See Groenzung *de ll. abrog. in Hollandia* ad tit. *arbor. furt. caesar.* l. 2. (One cruelly whipped at the Hague). See also Carpzovius in *Praxi Crim.*, part 2. quest. 83. num. 2. seqq. and several others: The German laws, concerning forests, are in abundance, and at large recited by Klochius and Pellerus.

the breaking up of those intervals, advance the growth of the trees to an incredible improvement.

2. It is therefore insisted on, that there is not a cheaper, easier or more prompt expedient to advance ship-timber, than to solicit, that in all his Majesty's forests, woods and parks, the spreading oak, &c. (which we have formerly described) be cherish'd, by plowing and sowing barley, rye, &c. (with due supply of culture and soil, between them) as far as may (without danger of the plowshare) be broken up. But this is only where these trees are arrived to some magnitude, and stand at competent distances; a hundred, or fifty yards (for their roots derive relief far beyond the reach of any boughs) as do the walnut-trees in Burgundy, which stand in their best plow'd-lands.

3. But, that we may particularize in his Majesty's Forests of Dean, Sherewood, Enfield-Chase, &c. and in some sort gratifie the quæries of the honourable the principal officers and commissioners of the Navy; I am advis'd by such as are every way judicious, and of long experience in those parts; that to enclose would be an excellent way : But it is to be considered, that the people, viz. foresters, and borderers, are not generally so civil and reasonable, as might be wished; and therefore to design a solid improvement in such places, his Majesty must assert his power, with a firm and high resolution to reduce these men to their due obedience, and to a necessity of submitting to their own and the publick utility, tho they preserved their industry this way, at a very tolerable rate upon that condition; while some person of trust and integrity did regulate and supervise the mounds and fences, and destine some portions frequently set apart for the

raising and propagating of wood, till the whole nation were furnish'd for posterity.

4. Which work if his Majesty shall resolve to accomplish, he will leave such an everlasting obligation on his people, and raise such a monument to his fame, as the ages for a thousand years to come, shall have cause to celebrate his precious memory, and his Royal successors to emulate his virtue. For thus (besides the future expectations) it would in present, be no deduction from his Majesty's treasure, but some increase, and fall in time to be a fair and worthy accession to it; whiles this kind of propriety would be the most likely expedient to civilize those wild and poor bordurers; and to secure the vast and spreading heart of the forest, which with all this indulgence, would be ample enough for a princely demesne: And if the difficulty be to find out who knows, or acknowledges what are the bordures; this article were worthy and becoming of as serious an inquisition, as the legislative power of the whole nation can contrive.

5. The sum of all, is; get the bordures well tenanted, by long terms, and easie rents, and this will invite and encourage takers; whilst the middle, most secure, and interior parts would be a Royal portion. Let his Majesty therefore admit of any willing adventurers in this vast circle for such enclosures in the precinct; and rather of more, than of few, though an hundred or two should join together for any enclosure of five hundred acres more or less; that multitudes being thus engaged, the consideration might procure and facilitate a full discovery of latter encroachments, and fortifie the recovery by favourable rents, improvements and reversions by copy-hold, or what

other tenures and services his Majesty shall please to accept of.

6. Now for the planting of woods in such places (which is the main design of this whole treatise) the hills, and rough grounds will do well ; but they are the rich fat vales and flats which do best deserve the charge of walls ; such as that spot affords ; and the haw-thorn well plash'd (single or double) is a better, and more natural fence, than unmorter'd walls, could our industry arrive to the making of such as we have describ'd : Besides, they are lasting and profitable ; and then one might allow sufficient bordure for a mound of any thickness, which may be the first charge, and well supported and rewarded by the culture of the land thus enclosed.

7. For example, suppose a man would take in 500 acres of good land, let the mounds be of the wildest ground, as fittest for wood : Two hedges with their vallations and trenches will be requisite in all the round, viz. one next to the enclosure, the other about the thicket to fence it from cattle : This, between the two hedges (of whatsoever breadth) is fittest for plantation : In these hedges might be tried the plantation of stocks, in the intervals all manner of wood-seeds sown (after competent plowings) as acorns, mast, fir, pine, nuts, &c. the first year chasing away the birds, because of the fir and pine seeds, for reasons given : The second year loosning the ground, and thinning the supernumeraries, &c. this is the most frugal way : Or by another method, the waste places of forests and woods (which by through experience is known and tried) might be perfectly cleansed ; and then allowing two or three plowings, well rooted stocks be set, cut and trimm'd as is requisite ; and

that the timber-trees may be excellent, those after-wards copp'ced, and the choicest stocks kept shreaded. If an enclosure be sowed, the seeds may be (as was directed) of all the species, not forgetting the best pines, fir, &c. Whiles the yearly removal of very incumbrances only, will repay the workmen, who fell the quick, or reserve it to store other enclosures, and soften the circumjacent grounds, to the very great improvement of what remains.

8. And how if in such fencing-works, we did sometimes imitate what Quintus Curtius, lib. 6. has recorded of the *Mardorum gens*, near to the confines of Hyrcania, who did by the close planting of trees alone upon the bordures, give so strange a check to the power of that great conqueror Alexander? They were a barbarous people indeed, but in this worthy our imitation ; and the work so handsomly, and particularly describ'd, that I shall not grieve to recite it. *Arbores densae sunt de industria consitae, quarum teneros adhuc ramos manu flectunt, quos intortos rursus in-serunt terrae : Inde, velut ex alia radice laetiores virent trunci : hos, qua natura fert, adolescere non sinunt ; quippe alium alii quasi nexu conserunt : qui ubi multa fronde vestiti sunt, operiunt terram. Itaque occulti ramorum velut laquei perpetuâ sepe iter claudunt, &c.* The trees (saith he) were planted so near and thick together of purpose, that when the boughs were yet young and flexible, bent and wreath'd within one another, their tops were bowed into the earth (as we submerge our layers) whence taking fresh roots, they shot up new stems, which not being permitted to grow as of them-selves they would have done, they so knit and perplex'd one within another, that when they were clad with leaves, they even covered the ground, and

enclosed the whole countrey with a kind of living
net, and impenetrable hedge, as the historian continues
the description ; and this is not unlike what I am told
is frequently practis'd in divers places of Devon ;
where the oaks being planted very near the foot of
those high mounds by which they separate their
lands, so root themselves into the bank, that when it
fails and crumbles down, the fence continues still
maintain'd by them with exceeding profit.   Such
works as these would become a Cato, or Varro indeed,
one that were *Pater Patriae, non sibi soli natus,* born
for posterity ; but we are commonly of another mould,

.... *& fruges consumere nati.*

9. A fair advance for speedy growth, and noble
trees (especially for walks and avenues) may be
assuredly expected from the graffing of young oaks
and elms with the best of their kinds; and where the
goodliest of these last are growing, the ground would
be plow'd and finely raked in the season when the
scales fall ; that the showers and dews fastning the
seed where the wind drives it, it may take root, and
hasten (as it will) to a sudden tree ; especially, if
seasonable shreading be apply'd, which has sometimes
made them arrive to the height of twelve foot by the
first three years, after which they grow amain.   And
if such were planted as near to one another as in the
examples we have alledged, it is almost incredible
what a paling they would be to our most expos'd
plantations, mounting up their wooden walls to the
clouds : And indeed the shelving and natural declivity
of the ground more or less to our unkind aspects, and
bleak winds, does best direct to the thickning of these

protections; and the benefit of that soon appears, and recompences our industry in the smoothness and integrity of the plantations so defended.

10. That great care be had of the seeds which we intend to sow has been already advised; for it has been seen, that woods of the same age, planted in the same soil, discover a visible difference in the timber and growth; and where this variety should happen, if not from the seed, will be hard to interpret; therefore let the place, soil and growth of such trees from whence you have your seeds, be diligently examin'd; and why not this, as well as in our care of animals for our breed and store?

11. As to the form, obey the natural site, and submit to the several guizes; but ever declining to enclose high-ways, and common-roads as much as possible. For the rest, be pleased to reflect on what we have already said, to encourage the planting of the large spreading oak above all that species; the amplitude of the distance which they require resigned to the care of the verderer for grazing cattle, deer, &c. and for the great and masculine beauty which a wild *quincunx*, as it were, of such trees would present to your eye.

12. But to advance the Royal forests to this height of perfection, I should again urge the removal of some of our most mischievously plac'd iron-mills; if that at least be true which some have affirmed, that we had better iron, and cheaper from foreigners, when those works were strangers amongst us. I am inform-ed, that the New-English (who are now become very numerous, and hindred in their advance and prospect of the continent by their surfeit of the woods which we want) did about twelve years since

begin to clear their high-ways by two iron-mills :
I am sure their zeal has sufficiently wasted our stately
woods, and steel in the bowels of their mother Old-
England; and 'twere now but expedient, their brethren
should hasten thither to supply us with iron for the
peace of our days; whilst his Majesty becomes the
great sovereign of the ocean, free commerce, *nemorum
vindex & instaurator magnus.*   This were the only
way to render both our countries habitable indeed,
and the fittest sacrifice for the royal oaks, and their
Hamadryads to whom they owe more than a slight
submission : And he that should deeply consider the
prodigious waste which these voracious iron and
glass-works have formerly made but in one county
alone, the county of Sussex, for 120 miles in length,
and thirty in breadth (for so wide and spacious was
the ancient *Andradswald*, of old one entire wood, but
of which there remains now little or no sign) would
be touched with no mean indignation: I named the
Sussex glass-works; but what spoil and prodigious
consumption the salt-works had made in Worcestershire, see the complaint of Mr. Camden speaking of
Feckenham Forest in his days, now necessitated to
use other coal; certainly, the goodly rivers and forests
of the other World, would much better become these
destructive works, our iron, and saw-mills, than these
exhausted countries ; and we prove gainers by the
timely removal: I have said this already, and I cannot
too often inculcate it for the concerns of a nation,
whose only protection (under God) are her wooden
walls.

13. Another thing to be recommended (and which
would prove no less than thirty years, in some places
forty, and generally twenty years advance) were a

good (if well executed) Act to save our standards, and bordering trees from the ax of the neighbourhood: And who would not preserve timber, when within so few years the price is almost quadrupl'd? I assure you standards of twenty, thirty, or forty years growth, are of a long day for the concernments of a nation.

14. And though we have in our general Chapter of Copp'ces, declar'd what by our laws, and common usage is expected at every fell (and which is indeed most requisite, till our store be otherwise supply'd) yet might much even of that rigor be abated, by no unfrugal permissions to take down more of the standards for the benefit of the under-woods (especially where, by over-dropping and shade they interrupt the kindly dews, rains, and influences which nourish them) provided that there were a proportionable number of timber-trees duly and throughly planted and preserved in the hedge-rows and bordures of our grounds; in which case, even the total clearing of some copp'ces would be to their great advance, as by sad experience has been taught some good husbands, whose necessities sometimes forced them to violate their standards, and more grown trees during the late tyranny.

15. Nor will it be here unseasonable to advise, that where trees are manifestly perceived to decay, they be marked out for the ax, that so the younger may come on for a supply; especially, where they are chiefly elms; because their successors hasten to their height and perfection in a competent time; but beginning once to grow sick of age, or other infirmity, suddenly impair, and lose much of their value yearly: besides, that the increase of this, and other speedy timber, would spare the more oak for navigation, and the sturdier uses.

How goodly a sight were it, if most of the demesnes of our countrey gentlemen were crown'd and
incircl'd with such stately rows of limes, firs, elms,
and other ample, shady and venerable trees as adorn
New-Hall in Essex, the seat of that Suffolk Knight
near Yarmouth, our neighbouring pastures at Barnes;
with what has been planted of later years by the
illustrious Marquess of Worcester ; the most accomplish'd Earl of Essex ; and even in less fertile soils,
though purer air at Euston, by the Right Honourable
the Earl of Arlington, Lord Chamberlain of his
Majesty's Houshold : and at Cornbury by the late
Lord Chancellor the Earl of Clarendon ; and is done,
nearer this imperial city, by the Earl of Danby, late
Lord High Treasurer of England, at Wimbleton ;
the noble Earl of Rochester (succeeding him in that
supreme office) at New-Park ; the Duke of Norfolk
at Albery, now the Lord Garnsey's ; Sir Robert
Cooke at Durdence, at Epsom ; now my Lord
Barkley's : At Bedington an ancient seat of the
Carews, famous for the first orange-trees planted in
the naked earth 100 years since, and still flourishing;
Row-hampton, Losely, Ashstead, seats, parks and
plantations ; the Earl of Devonshire's mores, Sir
Robert Howard, &c.

Besides what might have been seen (as to me they
were in perfection, and with admiration) the Royal
seats of Oatland, Richmond, and above all Nonsuch,
described by the judicious Camden, with deserved
eulogies.

All these, and more, in my own sweet county of
Surrey, inferior to none for pleasure and salubrity
of the air : To which we add the princely sojourns
of the adjoining county, Eltham and Greenwich, for

its park and prospect not only emulous, but in many respects exceeding that of the famous Thrasian Bosphorus from Constantinople : That palace namely at Greenwich, now turned into a stately and capacious colledge (the incomparable work of that accomplished architect Sir Chris. Wren) to which I had the honour to lay one of the first foundation stones, as the first treasurer of that Royal structure, erected for the reception and encouragement of emerited and well deserving sea-men and mariners, for its glorious fabrick, and conveniencies, exceeding any in Europe, dedicated to that excellent purpose. To these also belongs a park, as there did to that of Eltham. Nearer the metropolis yet are those of St. James's, Hide-Park, and that sweet villa (as now built, planted and embellish'd) of Kensington, deserving a particular description ; and for all that can be desirable of magnificence, Hamton-Court, truly great, in a most beautiful flat ; the palace, gardens, canale, walks, groves and parks ; the sweet and silent Thames gliding her silver streams to the triumphal Winsonian *Tempe*, raising its stately head, and which alone, has in view an hemisphere, as far as eyes and telescopes can distinguish earth from heaven : Thus from the keape, the terrace, parks and forests, equalling, nay exceeding any thing Europe can boast of.

Other sweet and delectable countrey seats and villa's of the nobless, rich and opulent citizens (about our Augusta) built and environ'd with parks, padocks, plantations, &c. adapted to country and rural seats, dispersed through the whole nation, conspicuous not only for the structure of their houses, built after the best rules of architecture ; but for situation, gardens, canals, walks, avenues, parks, forests, ponds, prospect and vistas, groves, woods, and large plantations, and other the most charming and delightful recesses, natural and artificial: But to enumerate and describe what were extraordinary in these and the rest, would furnish volumes : For who has not either seen, admired, or heard of,

Audly-End, Althorp, Awkland, Allington, Amphill, Astwell, Aldermaston ?

Bolsover, Badminton, Breckly, Burghly on the Hill, and the other Burghly ; Bockton, Buckhurst, Buckland, Bellroiro, Blechington, Bestwood, Broomhall ?

Castle-Rising, Castle-Ashby, Chatsworth, Charsley, Cornbery, Casiabery, Cobham, Cowdrey, Caversham, Cranburn-Park, Charlton, Copt-Hall, Claverton ? famous for Sir W. Basset's vine-yard, producing 40 hogsheads of wine yearly ? nor must I forget that of Deepden, planted by the Honourable Charles Howard of Norfolk, my worthy neighbour in Surrey.

Drayton, Dorington-park, Dean ?

Eastwell, Euston, Ecleswold, Edscomb, Easton, Eping ?

Falston, Flanckford ?

Graystock, Goodrick, Grooby, Grafton, Golden-Grove ?

Holdenby, Haddon, Hornby, Hatfeild, Haland, Hoathfield, Hinton, Holm-Pierpoint, Horstmounceaux?

Inchingfield ?

Kirby, Knowesby ?

Longleat, Latham, Lensdal, Latimer, Lawnsbourgh?

More-Park, Mulgrave, Marlborough ?

Normanby, North-hall, Norborough, Newnham ?

St. Ostlo, Oxnead ?

Petworth, Penshurst, Paston-Hall?
Quarendon, Quickswood?
Ragland, Rutford, Ragbey, Ricot?
Sherborn, Sherley, Swallowfield, Shasford, Shafts-
bery Stansted, Scots-hall, Sands of the Vine?
Theobalds, Thorn-kill, Thorny?
Up-Park?
Wilton, Wrest, Woburn, Welbeck, Worksop,
Woodstock, which as Camden tells us, was the first
park in England; as it is like to be one of the most
magnificent and princely palaces and seats of that
illustrious hero, his Grace, the Duke of Marlborough;
to whose courage and conduct not the safety of the
empire alone, but of Europe is due, whilst the actions
at Blenheim and Schellemberg may challenge equal
trophies with Miltiades and Cæsar, at Marathon and
Pharsalia. But to proceed Wimburn, Whittle-Park?

And generally all those seats which go under the
names of castles and halls, (as in Yorkshire, Essex,
&c.) were stor'd with noble parks full of timber,
omitted here; which, but to have nam'd, would
overswell the alphabet; without reckoning those of
Ireland, which few years since was an exhaustable
magazine of timber, destroyed by the Cromwellian
rebels, not only in that kingdom, but through all
England: As to parks, there were more in this nation,
than in all Europe beside: And most of all that catalogue above named, have yet their parks full of
good timber-trees, industriously improved by the
owners, since the spoil of the late usurpers and sequestrators.

To these should I add the vast forests, (most of
them belonging to the Crown) as that of Dean, New-
Forest, Windsor, Ashdown, Leonard, Sherwood, Ep-

ping, Panbet, Chute, &c. forests for the most part
without trees : And several of them together heretofore comprehended in that vast *Andradswald* already
mentioned, of one county only : There were formerly
twenty groves in Clarendon-Park near Salisbury, celebrated by Mesokerus, cited by Camden, that were
every one of them a mile in compass. In a word, to
give an instance of what store of woods and timber
of prodigious size, there were growing in our little
county of Surrey, (the nearest of any to London) and
plentifully furnished both for profit and pleasure,
(with sufficient grief and reluctancy I speak it) my
own grandfather had standing at Wotton, and about
that estate, timber, that now were worth 100000*l.*
Since of what was left my father, (who was a great
preserver of wood) there has been 3000*l.* worth of
timber fallen by the ax, and the fury of the late hurricane and storm : Now no more Wotton, stript and
naked, and ashamed almost to own its name.

All which considered (for there are many other
places and estates which have suffer'd the like calamity), should raise, methinks, a new spirit of industry in
the nobility and gentry of the whole nation, like that
which Nehemiah inspir'd the nobles, as well as the
people of the captivity[1] (than which nothing so much
resembled that tedious slavery, and return from it,
than did the restoration of King Charles II) Let us
arise up (says the brave man), and build ; and so they
strengthened their hand, for the people had a mind to
the work. And such an universal spirit and resolution,
to fall to planting, for the repairing of our wooden-
walls and castles, as well as of our estates, should truly
animate us : Let us arise then and plant, and not give

[1] Nehem. c. 2, v. 18.

it over till we have repaired the havock our barbarous enemies have made : Pardon then this zeal, O ye
lovers of your countrey, if it have transported me !
To you Princes, Dukes, Earls, Lords, Knights and
Gentlemen, noble patriots (as most concerned) I
speak, to encourage and animate a work so glorious,
so necessary: A spirit like this was that which so
universally excited, and set forward the repair of the
decay'd peer at Dover, built of timber ; gentlemen
and persons of all degrees, setting their hand to it,
with a wonderful and unanimous zeal and alacrity, as
it is described by our honest Holingshed, in the reign
of Q. Elizabeth.   And what has been done of later
date, in order to the improvement of their estates,
and ornament of their seats, we have already shew'd,
leading the way to those noble and honourable attempts,
the fruit of their hands and industry, in so few years,
already beginning to exalt their stately heads about
their estates and dwellings.

   To continue this then, let none be discouraged,
who have any generous regard to the good of their
country and posterity : Let us hear the Hessian bard,

[1] When either barren sands have kill'd the trees,
   Or diligent hewers fell'd them by degrees ;
   Then lest the earth should waste, and bare remain,
   They scatter seeds, and leave them on the plain:
   Hence to proceed, young stalkless leaves you'll find,
   Next slender stems, which with a stronger rind
   Invested, rise to trees: Of these is made

[1] Cum vel arena siti sterilis confecit iniquâ
Vel labor excidit diuturnus & arida facta est
Planities, tum ne jaceant loca vasta recisis
Arboribus, nova conficiunt, & semina mittunt
Sparsa solo vacuo, campisque injecta relinquunt :
Tum videas prodire novas sine stipite frondes,
Mox quoque cauliculos tenues, tum cortice robur

A youthful grove, yielding a lovely shade;
Until at last, their stately heads they rear,
And tall (as those which they succeed) appear,
Ready again the workmens tools to marr.
This various culture, by the Germans taught,
Most other nations into use have brought :
Such is the love of groves, that with delight,
Or ample profit may the pains requite.

Having before celebrated and described the famous
forest about Norimberg :

> [1] A wood with kind embraces, five [2] miles wide,
> Encompasses the town on every side.
> No whit inferior to th' Hercinian grove,
> Whether you profit most, or pleasure love.

Of which noble forest and privileges, such care has
been taken by many [3] emperors, that the very models
of the plows are still preserved, drawn by above an
hundred horse, when 200 years since, this Royal
plantation was begun, wisely presaging what ravage
might be made by the spoil which the wars have

Adnasci, parvosque umbram defendere ramos
Exiguam, teneramque novo de germine silvam
Surgere, & in patrias paulatim adolescere formas ;
Donec in antiquum redeat decus, altaque caelo
Attollat capita, & concusso vertice nutet
Lassatura iterum patrias jam silva secures,
Has aliæ innumeræ per tot jam secula, terræ
Rescivere artes reparandarum silvarum.
Inventrix docuit Germania, tanta cupido est
Tantus amor nemorum, quorum vel blanda voluptus,
Vel gravis utilitas sit responsura labori.

[1] Circuit inclusam pulchris amplexibus urbem
Silva patens passum per millia quinque recessu
Interiore sui, vel paulo plura, nec ulli
Herciniæ nemorum cedens, si commoda spectes ;
Aut etiam quæ silvarum solet esse voluptas,
Te juvet, atque animi tantum oblectamina quæras.

[2] German Miles in England. 20.
[3] Colerus Oecon. l. 8. c. I. II.

since caused in that goodly country; which being then an almost continual forest, is now so sadly wasted. Nor has this been the fate of Germany alone, but of all the most flourishing parts of Europe, thro' the execrable and unsatiable ambition of those who have been the occasion of the ruin not only of these venerable shades, stately trees and avenues, (the graceful ornaments of the most princely seats) but of the miserable desolation of entire provinces, which their legions have left, with the murders of so many Christians, inhumanly, and without distinction or just provocation! Mischiefs not to be repair'd in many ages, the truculent and savage marks (among others) of a most Christian King, *nomine non re !* In the mean time, what provision this demolisher of woods in other countries, makes to furnish and store his own dominions with so necessary a material, we have mention'd in this chapter, and how impolitick a waste there was of timber in France in John Bodins's time, see *Repub.* lib. VI. cap. I.

But (leaving this sad and melancholy prospect) I return to foreign descriptions (the effects of peace) and it shall be that plantation of elms, carried out of England by Philip the Second of Spain, to adorn his royal palace at Aranjuez (of which I have already spoken, cap. IV. lib. I.) near Madrid in Spain : The palace is seated on the bank of the famous river Tagus, and the plantation on the north, where there is a piece of ground inclos'd, form'd into walks of 680 yards long, and 300 in breadth, in shape of a trapezium or parallelogram, about which the Tago is artificially drawn to fence it. Next the river-side are more walks, not above 20 foot in breadth (for closer shade) planted on each side with double ranks

of elm, some of which are 40 yards high, stript up to
the top, and so near set, as 15 foot space : The second
row is about six foot distant from the other ; not
planted exactly against its usual opposite, but the
interval, and space, thro' which glides a narrow
shallow channel of water to refresh the trees upon
occasion ; thus,

```
o     o     o     o     o     o
_____

_____
   o     o     o     o     o
```

Which is the method us'd in many ridings of elm-
walks, some of which are a league in length, adorning
this seat beyond any palace (some think) in the world.
Many of these indeed are on the decay, prejudic'd by
their being planted so near one another : But for all
that, it takes not much from the beauty of the *vista*,
which is certainly the most surprizingly agreeable ;
to which the ample fountain, and noble statues in the
the cross-walks, make so glorious an addition, as
would require a particular description.

And now do I not for all this so magnify it, as if
not to be parallel'd in our own country ; where I dare
affirm, are many exceed it, both in form and planting,
(which has there several defects) but as we said, for
an exotick example, so admir'd and celebrated by
that boasting nation, as if the universe could not
shew the like.

And what, in the mean time, can be more delightful, than for noble persons, to adorn their goodly
mansions and demesnes with trees of venerable shade,
and profitable timber ? By all the rules and methods
imaginable, to cut and dispose those ampler enclosures

into lawns and ridings for exercise, health, and prospect, and for which I should here presume to furnish
some farther directions, were it not already done to
my hand by the often cited Mr. Cooke, in that useful
work of his ; where, in chapter the 38th. he has laid
down all that I can conceive necessary, by measures
exactly taken from the middle-line of any front,
following the center-stake, if it be for a walk : He
there determines the wideness of the walk, according
to its length, as 40 foot to one of half a mile ; if
more, 50 or 60 ; and if you withal desire shade, that
then you should make 3 walks, the two collaterals
20 foot broad, to a middle one of 40, 25 to 50, so
that the middle be as wide as both the other : He
likewise shews how proper it is that walks should
not terminate abruptly, but rather in some capacious
or pretty figure, be it circle, oval, semi-circle, triangle,
or square, especially in parks, or where they do not
lead into other walks ; and even in that case, that
there may gracefully be a circle to receive them :
There he shews how to pierce a walk through the
thickest wood either by stakes set up where they
may be seen to direct, or by candle and lantern, in a
calm night, &c.   He also gives the distances of the
trees in relation to each other, according to the
species, and shews how necessary it is, to plant them
nearer in those ovals, circles, and squares, &c. for
the better distinction of the figures, suppose to half
the distance of that of the walks, and proportionable
to the amplitude or smalness thereof : As for lawns,
he advises that they should (if possible) be contriv'd
on the south or east side of the seat and mansion, for
avoiding the impetuousness of western winds ; and
that your best rooms may front those lawns and

openings, and to skreen from the occidental and
after-noons sun, which also hinders prospect : A lawn
on the north, exposes the house to that piercing
quarter, and therefore it would be well defended with
the tallest trees : For the figure he commends the
square, with three avenues breaking out at the three
angles, or one at the angle opposite to the house ;
and these lawns may be bounded with walks, or a
single row of lime-trees at competent distance : To
which I add, the circle with a star of walks radiating
from it likewise exceeding pleasant ; such as the
Right Honourable the Earl of Winchelsea has cut
out at his noble Seat in Kent ; and since, (far exceeding the most) at Long Leats, the stately palace of the
Lord Viscount Weymouth ; at Badminton, the Duke
of Beaufort's princely Seat in Gloucestershire : At
Ackdowne Park in Berks, a most delightful solitude,
from the centre of a large wood belonging to the
Lord Craven ; and in Worcestershire at Westwood,
the mansion of Sir John Packington ; besides those
mention'd by Dr. Plot in his *Nat. Hist. of Staffordshire*,
with many others ; most of which have been graphic-
ally plotted and design'd, (together with the seats,
gardens, fountains, *piscinas*, plantations, avenues, *vista's*,
and prospects about them) by Mr. Kniff, in near an
hundred copper-plates : A most laudable undertaking,
and becoming the encouragement of those noble
persons, who would do honour to themselves, their
family, and whole nation. By these, and the like
examples, gentlemen (lovers of improvements) may
learn how to contrive and adapt a square, oblong,
regular or irregular figure, according as their woods,
groves and parks are dispos'd, and lie proper for
avenues and *vista's*, radiating from their Seats in the

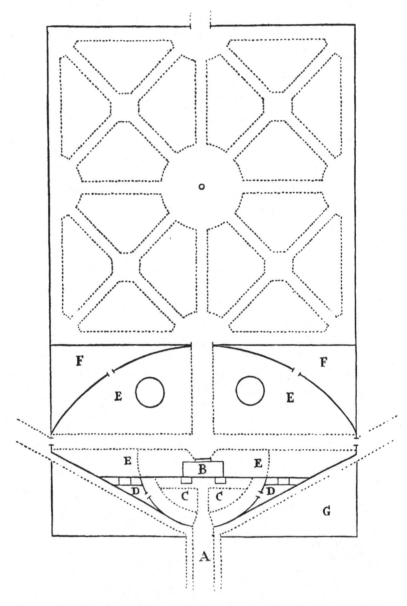

A, *The Principal* Avenue. B, *The* House. C, *The* Court *before it*.
D, *Place for* Stables, *and other* Offices.   E, Gardens *and* Fountain.
F, Hortyards *and* Fruit, &c.   G, Pasture.

country ; such especially, as are situated in spacious flats, or gentle declivities. For instance, suppose a large and spacious square, contiguous to the gardens, or some pasture-fields before it, (not here set out) were divided into several quarters, separated only by walks and ranks of trees, and if need be, inclos'd with low hedges of holly, yew, or other hardy greens: The trees and walks planted at competent distance with forest-trees, unmixt ; that is, what were planted with oak, should be set throughout with oak ; elm with elm, and so of the rest : That the walks which lead to or from the front of the house, be 100, or 150 foot in breadth, others 60, 50, &c. in proportion to the length, and the distance of the trees as the species require. But this is not so easily at first discern'd, by such as being desirous of speedy shade and ornament, plant their trees too near one another, which is a general error. Thus much concerning the walks and *vista's*. The vacuities, which are sixteen blunt triangles, and consider'd as pretty large fields, may be stored with several sorts of good timber-trees ; oak, elm, ash, walnut, beech, chesnut, lime, service, maple, black-cherry, fir, and pines, &c. some of them plow'd for corn, and left for meadow and pasture ; cyder, cherry, and other ortyard-fruit : Than which nothing could be more profitable and graceful.

I have omitted the basse-court, which may be added to the half circle C ; nor have I given the dimensions of any the separations or necessary buildings ; intending this as an idea only of something which I conceive might be both convenient and graceful, or to be varied into other figures, according to the pleasure of the owner. The black

ww

lines are walks ; the pointed, ranks of trees and walks.

And for an instance of irregular figures, actually survey'd, and dispos'd into walks, the following plot is presented to me by the ingenious Esq. Kirk, set out in a large wood of his (call'd Mosely) near his house at Cookeridge (betwixt Leeds and Oteley) in Yorkshire; the whole containing six-score acres : Nor are such glades thro' copp'ces to be neglected, in some regard preferable to the woods of taller trees, obnoxious to be subverted by impetuous storms, which the humbler copp'ces escape, and yet let in very noble views and prospects; besides their inviting of game for breed, and to shelter sonorous birds, which never are found in lofty woods, where they are expos'd to hawks and owles.

And here should I shut up this section, were I not most advantagiously as well as obligingly prevented, by the improvement following (sent me from the Reverend Mr. Walker), to shew how forest-trees may be planted in consort with fruit-trees, at once answering both profit and pleasure: Take it as himself describes it, which cannot be better.

' In open fields, where a man happens to have only
' single broad lands or leys lying by themselves, or
' only two or three lying together, in every such place
' he may set a row of trees near the middle, every
' second tree being a fruit-tree, and the rest forest-
' trees : Or, on narrow pieces never likely to be
' plowed (as on meadow ground, hades, &c.) betwixt
' two fruit-trees may be set two or more forest-trees,
' in a line crossing the row of fruit-trees, as in fig. 5th.
' On arable ground he may make balks, which may
' be mowed, and trees may be set on them. If upon

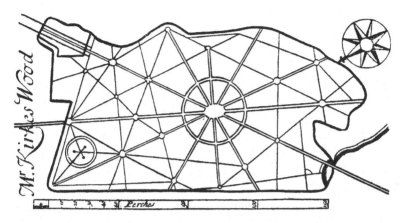

The lines in this *Platforme* represents the *Walkes* in *Mr. Kirke's Wood* (cal'd *Moseley*) neare his *House* at *Cookeridge*, (betwixt *Leeds* and *Otley*) in *York-Shire*. The whole containing about *Six Score Akers*.

The Double line *Walks* are about 20 *Foot* wide, and ye *Single* lines about 8. *Foot* wide.

Total of the Views	8	18	140	10	60	7	32	9	10	12	306
Number of Centers	4	6	35	2	10	1	4	1	1	1	65
Views	2	3	4	5	6	7	8	9	10	12	Sum

This *Table* shews in the first *Collumne* the *Number of Views* in each *Center*, in the *Second Collumne* ye *Number of Centers*, and in ye third ye total *Number* of all the *Views*.

' balks 4 foot broad, fruit-trees be set, 4 pole or 22
' yards asunder, and one forest-tree be set betwixt
' every two fruit-trees, then for every acre of ground
' left unplowed, there may be 160 fruit-trees, and 160
' forest-trees.

' Some think the loss of ground in making holes
' for trees and tumps, to be more than really it is :
' For, if a mark be paid yearly for an acre, this comes
' to no more than one penny for a square pole, which
' is 30 square yards and ¼; and he that pays 20s. for
' an acre, has 20 square yards for 1d.

' In closes, or on broad pieces of ground in open-
' fields, trees may be planted in some of the orders
' described in the 4 first figures following.

' In these figures each letter represents a tree, viz.
' a a a fruit-trees, 30 yards asunder in equilateral
' triangles ; and oo forest-trees.   In the 1st. fig. the
' rows are 15 yards asunder, the fruit-trees in the
' same row 52 yards asunder, the forest-trees 8 yards
' asunder, and 10 yards from the nearest fruit-trees.
' These forest-trees may be often pruned up to the
' top : The rows may run the same way that the lands
' or leys shoot.   In every acre about 6 fruit-trees,
' and 30 forest-trees may be thus planted : Or the
' distance may be more or less, as the planter thinks fit.

' In places never likely to be plowed, trees may be
' set as in the 2d. or 3d. fig.   In the 2d. fig. betwixt
' 3 fruit-trees are set 3 forest-trees, 8 yards asunder,
' and 15 yards and ½ from each fruit-tree.   A fruit-
' tree has 12 forest-trees round about it.   About 6
' fruit-trees, and 36 forest-trees may be thus set in
' one acre.

' In the 3d. fig. betwixt 3 fruit-trees are set 4
' forest-trees, 17 foot and ⅓ asunder.   Here round

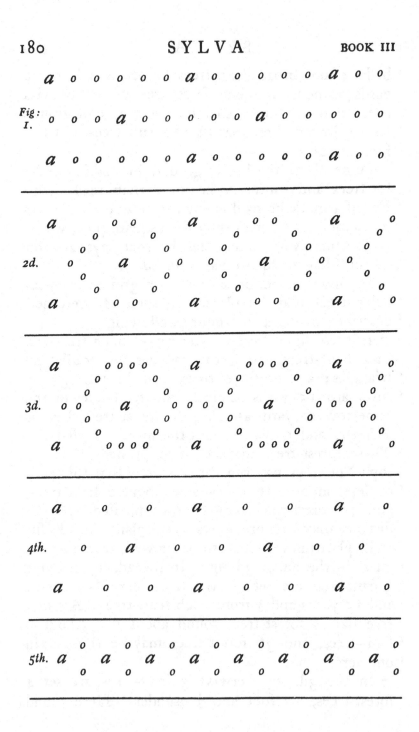

' about a fruit-tree stand 18 forest-trees, describing
' the figure of a hexagon, like one of the holes in a
' honey-comb. In each acre about 6 fruit-trees, and
' 48 forest-trees may be thus set.

   ' In the 4th. fig. all the trees are 17 yards and $\frac{1}{3}$
' asunder: Betwixt 3 fruit-trees stands 1 forest-tree.
' In each acre about 6 fruit-trees, and 12 forest-trees
' may be set thus.

And now to encourage this, gentlemen may not
only lawfully plant trees upon their own demesnes,
but in commons also, and open fields, in spacious
rows, or otherwise; provided they set them so far
from other mens grounds, as that their boughs hang
not over them (as we have shew'd was both by the
imperial, and our own constitution, prohibited) or no
nearer one another (in arable land) than such trees are
usually set in grounds inclos'd; that is, so as not to
hinder the plough. Such trees, if of fruit, so dispos'd
and set, belong intirely to the planter, (tithes excepted)
without that the commoner can challenge any part
thereof.

It would therefore be a most charitable work, to
plant fruit, and forest-trees for the benefit of the poor,
upon commons, and other waste grounds, and such
places where they would thrive; and where persons
are willing to give money to be thus employ'd for
the use of the indigent, among the sundry ways of
disposing of it to that end, as in the figures above
describ'd.

   16. But these incomparable amenities and under-
takings will best of all become the inspection and care
of the noble owners, lieutenants, rangers, and ingeni-
ious gentlemen, when they delight themselves as much
in the goodliness of their trees, as other men generally

do in their dogs, and horses, for races, and hunting; neither of which recreations is comparable to that of planting, either for virtue or pleasure, were things justly consider'd according to their true estimation : Not that I am of so morose an humour, that I reprove any of these noble and manly diversions, seasonably us'd ; but because I would court the industry of great and opulent persons, to profitable and permanent delights : For, suppose that ambition were chang'd into a laudable emulation, who should best, and with most artifice, raise a plantation of trees, that should have all the proper ornaments and perfections their nature is susceptible of, by their direction and encouragement ; such as Ælian sums up, lib. 3. c. 14. εὐγενεῖς οἱ κλάδοι, καὶ ἡ κόμη πολλὴ, &c. kind and gentle limbs, plenty of large leaves, an ample and fair body, profound, or spreading roots, strong against impetuous winds (for so I affect to read it) extensive and venerable shade, and the like : Methinks there were as much a subject of glory as could be fancied of the kind ; and comparable, I durst pronounce, preferable to any of their recreations ; and how goodly an ornament to their demesnes and dwellings, let their own eyes be the judges.

17. One encouragement more I would reinforce from an history I have read of a certain frugal, and most industrious Italian nobleman, who, after his lady was brought to bed of a daughter, (considering that wood and timber was a revenue coming on whilst the owners were asleep) commanded his servants immediately to plant in his lands (which were ample) oaks, ashes, and other profitable and marketable trees, to the number of an hundred thousand; as undoubtedly calculating, that each of those trees might be worth

twenty pence, before his daughter became marriageable, which would amount to 100000 Francs (which is near ten thousand pound sterling) intended to be given with his daughter for a portion. This was good philosophy, and such as I am assur'd was frequently practis'd in Flanders upon the very same account : Let us see it once take effect amongst our many slothful gentry, who have certainly as large demesnes, and yet are so deficient in that decent point of timely providing for their numerous children : And those who have none, let them the rather plant : Trees and vegetables have perpetuated some names longer, and better than a pedigree of a numerous off-spring (as I have already shew'd ;) and it were a pledge of a noble mind, to oblige the future age by our particular industry, and by a long lasting train, with the living work of our own hands. But I now proceed to more general concerns, in order to the quæries, and first to the proportion.

18. It were but just, and infinitely befitting the miserable needs of the whole nation, that every twenty acres of pasture made an allowance for half an acre of timber ; the ground dug about Christmas, casting the grassy side downwards till June, then dug again, and about November stirr'd afresh, and sown with mast, or planted in a clump, well preserv'd, and fenc'd for 14 or 15 years; unless that sheep might haply graze after 4 or 5 years : And where the young trees stand too thick, there to draw and transplant them in the hedge-rows, which would also prove excellent shelter for the cattle : This husbandry would more especially become Northamptonshire, Lincolnshire, Cornwall, and such other of our countries as are the most naked of timber, fuel, &c. and unprovided of

covert : For it is rightly observ'd, that the most fruitful places least abound in wood, and do most stand in
need of it.

1.

Example by Leicestershire,
   What soil can be better than that
For any thing heart can desire ?
   And yet doth it want ye see what :
Mast, covert, close pasture, and wood,
   And other things needful, as good.

2.

More plenty of mutton and beef,
   Corn, butter, and cheese of the best,
More wealth any where (to be brief)
   More people, more handsom, and prest,
Where find ye (go search any coast)
   Than there where inclosure is most ?

3.

More work for the labouring man,
   As well in the town as the field ;
Or thereof (devize if ye can)
   More profit what countries do yield ?
More seldom where see ye the poor
   Go begging from door to door ?

4.

In wood-land the poor men that have
   Scarce fully two acres of land,
More merrily live, and do save
   Than t'other with twenty in hand :
Yet pay they as much for the two
   As t'other for twenty must do.
   If this same be true, as it is,
   Why gather they nothing by this ?

Thus honest Tusser above an hundred years since, and the whole age has justified it; since 'tis evident, that by inclosure and this diligent culture, the very worst land of England would yield tenfold more profit, than that which is here celebrated for the best and richest spot of it.

19. Such as are ready to tell ye their lands are so wet, that their woods do not thrive in them, let them be converted to pasture; or bestow the same industry on them which good husbands do in meadows by draining; which instead of those narrow rills (and gutters rather) might be reduc'd to a proportionable *canale* cut even and strait; the earth taken out, spread upon the weeping and uliginous places : Nor would the charge be so much, as that of the yearly and perpetual renewing, and cleansing of those numerous and irregular slices ; beside the profit of storing the canal with fish.

It is a slothfulness to do otherwise, since it might be effected in few years, by continually, and by degrees making the middle cut large, where it cannot be so conveniently done at once, and the pains would certainly be as fully recompenc'd in the growth of their timber, as in that of their grass: Where poor hungry woods grow, rich corn, and good cattle would be more plentifully bred; and it were beneficial to convert some wood-land (where the proper vertue is exhausted) to pasture and tillage; provided, that fresh land were improved also to wood in recompence, and to balance the other.

20. Where we find such uliginous and starv'd places (which sometimes obey no art or industry to drain, and of which our pale and fading corn is a sure indication) we are as it were courted to obey nature,

and improve them from the propagation of sallows,
willows, alders, abele, black-cherry, sycomore, aspene,
birch, and the like hasty and profitable growers, by
ranging them, casting of ditches, trenches, &c. as
before has been taught.

21. In the mean while, 'tis a thing to be deplor'd,
that some persons bestow more in grubbing, and
dressing a few acres which have been excellent wood,
to convert them into wretched pasture, not worth a
quarter of what the trees would have yielded, well
order'd, and left standing ; since it is certain, that
barren land planted with wood, will trebble the expence in a short time.   Of this, the Right Honourable
the Lord Viscount Scudamor may give fair proof,
who having fell'd (as I am credibly inform'd) a
decay'd wood, intended to be let to tenants ; but upon
second thoughts, (and for that his Lordship saw it
apt to cast wood) enclos'd and preserv'd ; it yielded
him, before thirty years were expir'd, near 1000
pound upon wood-falls ; whereas the utmost rent of
the whole price of land yearly, was not above 8 pound
10 shillings.   The like I am able to confirm by in-
stancing a noble person, who (a little before our unhappy wars) having sown three or four acres with
acorns, the fourth year transplanted them which
grew too thick all about his lordship : These trees
are now of that stature, and so likely to prove excellent timber, that they are already judg'd to be
almost as much worth as the whole demesnes ; and
yet they take off nothing from other profits, having
been discreetly dispos'd of at the first designment.
And supposing the longaevity of trees should not
extend to the periods we have (upon so good account)
produc'd ; yet, neither is their arrival to a very com-

petent perfection, so very discouraging ; since I am
credibly inform'd, that several persons have built of
timber (and that of oak) which were acorns within
this forty years ; and I find it credibly reported, that
even our famous Forest of Dean, hath been utterly
wasted no less than three several times, within the
space of nine hundred years. The Prince Elector
Frederic IV. in the year 1606 sow'd a part of that
most barren heath of Lambertheim, with acorns after
plowing, as I have been inform'd : It is now likely
to prove a most goodly forest, though all this while
miserably neglected by reason of the wars. For the
care of planting trees, should indeed be recommended
to princes and great persons, who have the fee of
the estate ; tenants upon the rack, by reason of the
tedious expectation, and jealousie of having their rents
enhanc'd, are for the most part averse from this husbandry ; so that unless the landlord will be at the
whole charge of planting and fencing, (without which
as good no planting) little is to be expected ; and
whatsoever is propos'd to them above their usual
course, is look'd upon as the whim and fancy of spe-
culative persons, which they turn into ridicule when
they are applied to action ; and this, (says an ingenious
and excellent husband, whose observations have afforded me no little treasure) might be the reason,
why the prime writers of all ages, endeavour'd to
involve their discourses with allegories, and aenig-
matical terms, to protect them from the contempt
and pollution of the vulgar, which has been of some
ill consequence in husbandry ; for that very few
writers of worth have adventured upon so plain a
subject ; though doubtless to any considering person,
the most delightful kind of natural philosophy, and

that which employs the most useful part of the mathematics.

The Right Honourable the late Lord Viscount Mountague has planted many thousands of oaks, which I am told, he drew out of copp'ces, big enough to defend themselves ; and that with such success, as has exceedingly improv'd his possessions ; and it is a worthy example. To conclude, I could have shewn an avenue planted to a house standing in a barren park, the soil a cold clay ; it consisted totally of oaks, one hundred in number : The person who first set them (dying very lately) lived to see them spread their branches 123 foot in compass, which at distance of 24 foot, mingling their shady tresses for above 1000 in length, form'd themselves into one of the most venerable and stately arbor-walks, that in my life I ever beheld : This was at Baynards in Surrey, and belonging lately to my most honour'd brother, (a most industrious planter of wood) Richard Evelyn, Esq ; since transplanted to a better world : The walk is broad 56 foot, and one tree with another, containing by estimation three quarters of a load of timber in each tree, and in their lops three cords of fire-wood : Their bodies were not of the tallest, having been topped when they were young, to reduce them to an uniform height ; yet was the timber most excellent for its scantling, and for their heads, few in England excelling them : Where some of their contemporaries were planted single in the park without cumber, they spread above fourscore foot in arms ; all of them since cut down and destroy'd, by the person who continued to detain the just possession of that estate, from those to whom of right and conscience it belong'd. Since then it is dispos'd of, I am glad it is fallen into the hands of the present possessor.

22. But I have some few instances to superadd, of no mean encouragement, before I dismiss my reader, because they are so very pregnant and authentick. Sir Tho. Southwel, after he had sold, and fell'd all the timber and under-wood in a certain parcel of land lying in Carbrook, in the County of Norfolk, call'd by the name of Latimer Wood, containing 80 acres (now, as I understand, belonging to Sir Rob. Clayton, Knight) granted a lease of the said ground, with other land, to one Tho. Wastney (the father) with liberty to grub and stub-up all the wood and stub-shoots remaining, and to clear the said ground for pasture or tillage, as he should think to be most for his profit and advantage : Accordingly he puts out the same to labourers to stub and clear ; but was, it seems, perswaded by one of them, to preserve some of the young stands or saplings then growing there, as that which might be of greater emolument to him before the expiration of the lease, than if he should quite extirpate them, and convert the said ground to tillage : These saplings were then so small, as when it happen'd that any of the labourers did break the haft of his mattock, he could hardly find one amongst them, big enough to make another of for his present use : Nay, when the said labourers had made an end of clearing the ground of the old stub-shoots, upon which the timber and under-wood did grow (which is now 50 years since) there was not a tree left growing in it, that could be valued at above three pence to be fell'd for any use or service : About the year 1650, the estate being then come (after the death of Sir Rich. Crane, Knight) to

William Crane, Esq; and the lease of the same to
Tho. Wastney (the son) he offered 500 of the best of
the said young oak-sapplings to one Daniel Hall (a
dealer in timber) for two shillings and six pence the
tree, which he refusing to give, the said Tho. Wastney,
making his application to Mr. Crane above-mention'd
(then owner of the estate) and desiring Daniel Hall
to acquaint him what pity it was to cut down such
young and thriving trees ; Mr. Crane was perswaded
to allow the said Tho. Wastney fourscore pounds,
and to let them stand; since which time, the said
Mr. Crane sold as many of those trees and saplings,
as came to about forty pounds, and left growing, and
remaining on the ground about 1380 trees ; which,
in August 1675, being (upon the desire of Mr. Crane)
valued by the said Daniel Hall, were estimated to be
worth 700*l.* himself since offering for some of the
said trees 40 and 50 shillings a tree; 500 of them
being better worth than 500*l.*   Now the said Latimer
Wood were it clear'd of the timber, would not be let
for above four or five shillings per acre at the most.
The particulars of this history I received under the
hands and certificates of the above-mention'd Daniel
Hall, who is the timber-merchant, and two of the
stubbers or labourers (yet living) that were employ'd
to clear the ground.   I have likewise transmitted to
me this account from Mr. Sharp, under the hand of
Robert Daye, Esq; one of his Majesty's Justices of
the Peace for the county of Norfolk, as followeth.

   There were in 1636 an hundred timber-trees of
oak, growing on some grounds belonging then to
Thomas Day of Scopleton, in the county of Norfolk,
Esq ; which were that year sold to one Rob. Bowgeon
of Hingham in the said county, for 100 *l.* which

price was believed to equal, if not to surmount their intrinsick worth and value; for, after agreement made for them, a refusal happening (which continu'd the trees standing till the year 1671) those very trees were sold to Tho. Ellys of Windham (timber-master) and one Hen. Morley, carpenter, by Mr. Day (son of the said Thomas Day, Esq ;) for 560 pounds : And this comes to me attested under the hand of Esquire Day himself, dated 4 May 1678.

From the same Mr. Sharp I receive this instance of an ash planted by the hands of one Mr. Edm. Salter in that county, which he sold for 40s. before his death ; but this is frequent.

I am likewise assur'd that three acres of barren land, sown with acorns about 60 years since, and now become a very thriving wood, the improvement of those few acres amounts to 300 l. more than the rent of the land, and what it was before worth to be sold: Once more, and I have done.

Upon the estate of George Pitt, Esq ; of Strat-feildsea, in the county of Southampton, a survey of timber being taken in the year 1659, it came to 10300 l. besides near 10000 samplers not valu'd, and growing up naturally : Since this, there hath been made by several sales 5600 l. and there has been fell'd for repairs, building and necessary uses to the value, (at the least) of 1200 l. so as the whole falls of timber amount to 6800 l. The timber upon the same ground being again survey'd anno 1677, appears to be worth above 21000 l. besides 8 or 9000 samplers, and young trees to be left standing, and not reckon'd in the survey : But what is yet to be observed, most of this timber above-mention'd, being oak, grows in hedg-rows, and so as that the standing of it does very little prejudice to the plow or pasture.

It is likewise affirm'd, that upon a living in the same place, of about 40 *l. per an.* rent, there was (by an estimation taken in the year 1653) three hundred thirty eight young timber trees valu'd at fifty nine pound ; the saplings at thirty one pound fourteen shillings : And upon a later survey taken the last year 1677, the worth of the timber on that living, is valued at above eight hundred pound, besides four or five hundred young thriving trees, which have since the survey in 1653 grown naturally up, not reckoned in this account. With such, and the like instances, coming to me from persons and gentlemen of unquestionable credit (dispersed through several other counties of this nation) I might furnish a just volume ; and I have produced these examples, because they are conspicuous, full of encouragement, worthy our imitation ; and that from these, and sundry others which I might enumerate, we have made this observation, that almost any soil is proper for some profitable timber-trees or other, which is good for very little else.

23. Besides common pasture which has long been fed, and is the very best, meadow, that is up-land and rich, and such as we find to be naturally wood-seere (as they term it) the bottoms of downs, and like places well plow'd and sown, will bear lusty timber, being broken up, and let lie till Midsummer, and then stirred again before sowing about November.

Mr. Cook's directions are these : Prepare as for sowing of barly, about February scatter your seeds : If you plow your ground into great ridges, the thickness of the earth on the top will afford more depth and nourishment for the roots, and the furrows being filled up with leaves, when rotten, will lead the roots from one ridge to another : In dry ground plow the ridges cross the descent, not to drain, but keep the water on the ground, but in wet lands, contrary : This I hold to be an excellent note : He conceives the barly season to be of the latest to sow your seeds, but with oats it does well, so you sow them not too thick ; but 'tis best of all to sow them by themselves, without any crop of grain at all.

A more expeditious way is to plant with sets, making holes or fosses (which are best) two foot wide, and deep, and about half a rod distant, *viz.* four in every rod square, two sets in each hole, sowing your keys and seeds among them the ensuing Spring, and that continued as oft as you find stampings and keys to be had, even till your wood be perfectly furnished, only taking care that they lie not long too thick, because it will heat and burn the kernels, and therefore let them be put into the ground as soon as they are press'd or else lay them thin or parted with straw.

In case your land be poor, and wanting depth, or but indifferent, observing the posture of your ground, divide it into four yards distance at both extreams, by small stakes, making rows of them by setting up some few between them to direct, and lay your work straight, ploughing one yard of each side of the stakes, if the ground be green-sward for the easier running of the roots: Having thus ploughed two yards, and left two unploughed through your whole piece some short time before planting season, so soon as the fall of the leaf begins, dig up the unplough'd interstices, laying one half of the earth on the un-

plow'd pieces, and the other half upon the rest, and
as you do this, plant your prepared sets about a yard
distant, with store of sallow, or other cuttings with
them, digging that ground which you laid on the
plowed part a good spade deep, which will make it
near a foot thick to plant your sets in : Thus proceed
from one unplow'd ground to another till all of it is
planted: Two men on each side of the ridges will
soon dispatch the work, which would be finished by
the latter end of January, which is the best time for
the sowing your keys, nuts, and other seeds, unless
the weather be frosty, in which case you may a little
defer it : And when all is sow'd, cover them a little
with the shovelings of some ditches, pond, or other
stuff, as an assured good way to improve such grounds
to considerable advantage.

For the planting of wallnuts, chesnuts, cider-apples
or any other forest or fruit-tree, in open fields, Mr.
Cook directs how the triangular form exceeds all the
rest for beauty and advantage : I refer you to his 33
Chap.

An old and judicious planter of woods, prescribes
us these directions, for improving of sheep-walks,
downs, heaths, &c. Suppose, on every such walk on
which 500 sheep might be kept, there were plow'd
up twenty acres (plow'd pretty deep, that the roots
might take hold, and be able to resist the winds) this
should be sowed with mast of oak, beech, chats of
ash, maple-keys, sloes, service-berries, nuts, bullis, &c.
bruis'd crabs and haws, mingled and scattered about
the sides and ends of the ground, near a yard in
breadth. On the rest sow no haws, but some few crab-
kernels : Then begin at a side, and sow five yards
broad, plowing under the mast, &c. very shallow ;

then leave six yards in breadth, and sow and plow
five yards more, and so from side to side, remembring
to leave a yard and half at the last side ; let the rest
of the head-lands lie, till the remainder of the close
be sown in March with oats, &c. to preserve it from
hurt of cattel, and potching the ground ; when the
spring is of two years growth, draw part of it for
quick-sets ; and when the rest of the trees are of six
years shoot, exhaust it of more, and leave not above
forty of either side, each row five yards distant ; and
here and there a crab-stock to graff on, and in the
invironing hedge (to be left thick) let the trees stand
four yards asunder ; which if forty four were spared,
will amount to about 4000 trees: At twenty years end
stock up 2000 of them, lop a thousand more every
ten years, and reserve the remaining thousand for
timber : Judge what this may be worth in a short
time, besides the grass, &c. which will grow the first
six or seven years, and the benefit of shelter for sheep
in ill weather, when they cannot be folded ; and the
pasture which will be had under the trees, now at
eleven yards interval, by reason of the stocking up
those 2000 we mentioned, excepting the hedges; and
if in any of these places any considerable waters fortune to lie in their bottoms, fowl would abundantly
both breed and harbour there.   These are admirable
directions for park-lands, where shelter and food is
scarcy.

   But even this improvement yet does no way reach
what I have met withal in the most accurate, and no
less laborious calculation of Captain Smith upon this
very topic ; where he demonstratively asserts, that a
thousand acres of land, planted at one foot interval
in 7201 rows, taking up 51854401 plants of oak,

ash, chesnut, (or to be sown) taking up 17284800 of each sort, and fit to be transplanted at three years period (if set in good ground) are worth eighteen pence the hundred; and there being 345696 hundred, it amounts to no less than 25927*l*. 4*s*. besides the chesnuts, of which there being 1728480. (valued at, and worth half a crown the hundred) they come to 21606*l*. and the total of all, to 47533*l*. 4*s*.

This being made out, consider what an immense sum great trees would amount to, and in a large quantity of land; such as were worthy a Royal undertaking: It is computed, that at three foot distance, the first felling (that is, eight or nine years after their planting) would be worth in hoops, poles, firing, &c. 55015*l*. and the second fell, 28657*l*. 19*s*. 5*d*. And the fourth (which may be about thirty two years from their semination) 90104*l*. 17*s*. and so forward.

At four foot interval, and felling according to the same proportion, you may likewise reckon; and in 11 years, with 3 years crop of wheat (sow'd at first between) it will amount to 34001*l*. 9*s*. 4*d*. and the next, very much more; in regard the wood will spring up thicker: So as at the fifth fell, the account stands 126992*l*. 10*s*. 2*d*. &c. and at the seventh (whoever lives to it) 200000: And if planted at wider distance, *viz*. 18 foot (according to the Captain's method) at 30 or 40 years growth, you may compute them worth 192961*l*. 6*s*. and in seventy years, 201001; besides the three years crop of wheat; in all 410312*l*. 16*s*. which at 36 foot interval (accounted the utmost for timber) takes up (for 1000 acres) 40401 trees for the first 100 years. Then,

To make room, as they grow larger, grubbing up every middle tree, at 9*l*. per tree, 19800 trees amount to 99000*l.* and the remaining 20601 at 220 years growth, at but 8*l.* per tree, comes to 164808*l.* besides the inferior crop of meadow, or corn in all this time, sown in the distances; reckoning for three years product 90000 bushels at 5*s. per* bushel, which will amount to 22500*l.* besides the straw, chaff, &c. which at 5*s.* a load, and 3*d.* a bushel chaff, comes to 2025*l.* So as the total improvement (besides the 217 years emolument arising from the corn, cattel, &c.) amounts to 288333*l.*

And these trees (as well they may) coming to be worth for timber, 20*l.* an oak ; the 20601 trees amount to 412020*l.* and the total improvement of the 1000 acres (the corn profits not computed) ascends to 675833*l.* So as admit there were in all England (and which his Majesty might easily compass, even for his own proportion, and for posterity) 20000 acres thus planted, at two foot diameter (and, as may be presum'd, thirty foot high, which in 150 years they might well arrive to) they would be worth 13516660*l.* an immense and stupendous sum, and an everlasting supply for all the uses both of sea and land : But it is to Captain Smith's laborious works (to which I wish all encouragement) that we have the total charge of this noble undertaking from the first semination, to their maturity ; by which it will be easie to compute what the gains will be for any greater or lesser quantity.

But now to return to the place of planting (from whence this calculation has more than a little diverted) we shall find, as we said, that even in the most craggy, uneven, cold and exposed places, not fit for arable, as in Biscay, &c. and in our very peaks of Derbyshire, and other rocky places, ashes grow about every village, and we find that oak, beech, elm and ash will prosper

in the most flinty soils. And it is truly from these
indications, more than from any other whatsoever,
that a broken and decaying farmer is to be distinguish'd
from a substantial free-holder, the very trees speaking
the condition of the master : Let not then the Royal
patrimony bear a bankrupt's reproach : But to descend
yet lower ;

24. Had every acre but three or four trees, and as
many of fruit in it as would a little adorn the hedge-
rows, the improvement would be of fair advantage in
a few years; for it is a shame that turnip-planters should
demolish and undoe hedge-rows near London, where
the mounds and fences are stripp'd naked, to give sun
to a few miserable roots, which would thrive altogether
as well under them, being skilfully prun'd and lopp'd :
Our gardeners will not believe me, but I know it to
be true, tho Pliny had not affirmed it : As for elms
(saith he) their shade is so gentle and benign, that it
nourishes whatsoever grows under it : And (lib. 17.
c. 22.) it is his opinion of all other trees (very few
excepted) provided their branches be par'd away, which
being discreetly done, improves the timber, as we
have already shewed.

Indeed where elms are planted either about very
small crofts or avenues referv'd for pasture, the roots
are apt to spring up and annoy the grass : But I speak
of the larger field, and even in the former, that part
of the root which spreads into the field, may (as I
have shewn) be hinder'd from infecting it, by cutting
away those fibers which run into the field, without any
impeachment to the growth of the trees ; of which I
have some whose roots are cut off very near the main
stems at one side, thriving almost altogether as well
as those which have their roots entire.

25. Now let us calculate a little at adventure, and much within what is both fasible, and very possible ; and we shall find, that four fruit-trees in each acre throughout England, the product sold but at six-pence the bushel (but where do we now buy them so cheap ?) will be worth a million yearly : What then may we reasonably judge of timber, admit but at the growth of four pence *per* acre yearly (which is the lowest that can be estimated) it amounting to near half a million? if (as 'tis suppos'd) there may be five or six and twenty millions of square acres in the kingdom (besides fens, high-ways, rivers, &c. not counted) and without reckoning in the mast, or loppings; which whosoever shall calculate from the annual revenue, the mast only of Westphalia (a small and wretched country in Germany) does yield to that prince, will conclude to be no despicable improvement.

26. In this poor territory, every farmer does by ancient custom plant so many oaks about his farm, as may suffice to feed his swine : To effect this, they have been so careful, that when of late years the armies infested the poor country, both Imperialists, and Pro-testants ; the only bishoprick of Munster was able to pay one hundred thousand crowns *per mensem* (which amounts of our money to about 25000*l.* sterling) besides the ordinary entertainment of their own Princes and private families.   This being incredible to be practis'd in so extream barren a country, I thought fit to mention, either to encourage, or reproach us : General Melander was wont to say, the good husbandry of their ancestors had left them this stock *pro sacra anchorâ*; considering how the people were afterward reduc'd to live even on their trees, when the soldiers had devoured their hogs ; redeeming themselves from

great extremities, by the timber which they were at last compell'd to cut down, and which, had it continued, would have prov'd the utter desolation of that whole country.

I have this instance from my most worthy and honourable friend Sir William Cursius (late his Majesty's Resident in Germany) who received this particular from the mouth of Melander himself: In like manner, the Princes and Freedoms of Hesse, Saxony, Thuringia, and divers other places there, make vast incomes of their forest-fruit (besides the timber) for swine only : So as in a certain wood in Hassia only, twenty thousand have been fatted, yielding the Prince 30000 florins.

I say then, whosoever shall duly consider this, will find planting of wood to be no contemptible addition, besides the pasture much improved, the cooling of fat and heavy cattle, keeping them from injurious motions, disturbance, and running as they do in summer, to find shelter from the heat and vexation of flies.

27. But I have done, and it is now time to get out of the wood, and to recommend this, and all that we have propos'd, to his most sacred Majesty, the honourable Parliament, and to the Lord High Treasurer, principal officers, and commissioners of the Royal Navy; that where such improvements may be made, it be speedily and vigorously prosecuted; and where any defects appear, they may be duly reformed.

28. And what if for this purpose there were yet some additional office constituted, which should have a more universal inspection, and the charge of all the woods and forests in his Majesty's dominions? This

might easily be performed by deputies in every county ; persons judicious and skilful in husbandry ; and who might be repair'd to for advice and direction; And if such there are at present (as indeed our laws seem to provide) that their power be sufficiently amplified where any thing appears deficient; and as their zeal excited by worthy encouragements, so might neglects be encounter'd by a vigilant and industrious check.   It should belong to their province, to see that such proportions of timber, &c. were planted and set out upon every hundred, or more of acres, as the Honourable Commissioners have suggested ; or as might be thought convenient, the quality and nature of the places prudently considered : It should be their office also to take notice of the growth and decay of woods, and of their fitness for publick uses and sale, and of all these to give advertisements, that all defect in their ill governing may be speedily remedied ; and the superior officer or surveyor, should be accountable to the Lord Treasurer, and to the principal officers of his Majesty's Navy for the time being : And why might not such a regulation be worthy the establishing by some solemn and publick Act of State, becoming our glorious Prince, SOVEREIGN OF THE SEAS ; and his prudent senate, this present Parliament ?

But to shew how this *xylotrophiae studium* for the preservation of timber was honour'd,

29. We find in Aristotle's *Politics*, [1] the constitution

---

[1] De collegiis fabrorum, centonariorum,& dendrophororum, naviculariae.,ratium exercitor., & caudicariorum, plurimae extant inscriptiones apud Lipsium in lib. inscrip. antiq. quales Bergomensium, Brixianor., Comensium, Lugdunens., Avaricorum & Rhodanicor. eorumque corporum, & collegiorum patronis curatoribus. Vide etiam Hieron. Rubeum lib. 1. Hist. Ravennat. Item de Dendrophoris Lod. Theodos. lib. 1. & 2. iisdem verbis inscripto : Morisot. Orb. Marit. lib. 1. cap. 24.

of extra-urban magistrates to be *silvarum custodes;* and such were the *consulares silvae,* which the great Caesar himself (even in a time when Italy did abound in timber) instituted; and was one of the very first things which he did, at the settling of that vast empire, after the civil wars had exceedingly wasted the country : Suetonius relates it in the life of Julius; and Peter Crinitus in his fifth book *De honesta disciplina,* c. 3. gives this reason for it, *ut materies* (saith he) *non deesset, qua videlicet navigia publica possent a praefecturis fabrum confici:* True it is, that this office was sometimes called *Provincia minor ;* but for the most part, annex'd and joined to some of the greatest consuls themselves; that facetious sarcasm of the comoedian (where Plautus names it *provincia caudicaria*) referring only to some under officer, subservient to the other : And such a charge is at this day extant amongst the noble Venetians, who have near Trivisi (besides what they nourish in other places) a goodly forest of oaks, preserved as a jewel, for the only use of the arsenal, called the Montello, which is in length twelve miles, large five, and near twenty miles in compass; carefully supervised by a certain officer, whom they name *il capitano:* The like have the Genoëzes for the care of the goodly forests of Aitonae, in the Island of Corsica, full of goodly oaks and other timber ; which not only furnish that state with sufficient materials to build their own gallies and other vessels, but so many for sale to other nations, that since the late insult the French made upon that glorious city, he has haughtily forbid them to traffick any more with strangers, by supplying them as heretofore, to their great detriment and loss : This timber is of such a grain and quality, as though felled in the new-moon, it is not at all impair'd.

We might, besides all these, instance in many
other prudent states ; not to importune you with the
express laws which Ancus Martius the nephew of
Numa and other princes long before Caesar, did
ordain for this very purpose; since indeed, the care
of so publick and honourable an enterprize as is this
of planting and improving of woods, is a right noble
and royal undertaking ; as that of the Forest of Dean;
&c. in particular (were it bravely manag'd) an
Imperial design ; and I do pronounce it more worthy
of a prince, who truly consults his glory in the
highest of his subjects, than that of gaining battels,
or subduing a province.

And now after all this, and the directions and
encouragements enumerated in this chapter, together
with the most important concerns of these dominions,
and (next to God's immediate protection) the only
and most necessary expedient to preserve them : By
whose negligence so little effects appear of these
improvements which might by this time have been
made in the Royal Magazines ever since the first
edition of this treatise (and of so fair a growth of
useful timber) I list not to declare ; though the
officers then intrusted, and whose duty it was, be
now no more : I cannot, however, but call to mind
how seemingly solicitous and earnest the commissioners were, I should digest and methodize the
papers I laid before them on this subject, with a zeal
becoming publick spirits (as under their hands I have
to shew) whilst the putting it in practice to any
laudable degree, was soon cast by as a project scarce
worth the while. I again affirm, that had these
advantages of forest culture been then vigorously
encouraged and promoted, there had now been of

those materials infinite store, even from the very acorn and seminary, a competent advance of the most useful timber for the building of ships, (as I think is sufficiently made out) since his late Majesty's Restoration : The want of timber, and the necessity of being supply'd by foreign countries, if not prevented by better and more industrious instruments, may prove in a short time a greater mischief to the publick, than the late diminution of the coin.   I wish I prove no prophet, whilst I cannot for my life but often think of what the learned Melanchthon above a hundred years since was wont to say (long before those barbarous wars had made these devastations in Germany), that the time was coming, when the want of three things would be the ruin of Europe, *lignum, probam monetam, probos amicos ;* timber, good money, and sincere friends : How far we see this prediction already verify'd, let others judge : And if what I here have touch'd with some resentment in behalf of the publick and my country, in this rustick discourse, and us'd the freedom of a plain forester, seems too rude ; it is the person I was commanded to put on, and my plea is ready,

Δρυὸς παρούσης, πᾶς ἀνὴρ ξυλεύεται.

Praesente quercu, ligna quivis colligit.

For who could have spoken less upon so ample a subject ? and therefore I hope my zeal for it in these papers, will excuse the prolixity of this digression, and all other the imperfections of my services.

Si canimus silvas, silvae sunt consule dignae.

DENDROLOGIA

THE FOURTH BOOK

An historical account of the sacredness and use of standing groves, &c.

1. And thus have we finish'd what we esteem'd necessary for the direction of planting, and the culture of trees and woods in general; whether for the raising of new, or preservation of the more ancient and venerable shades, crowning the brows of lofty hills, or furnishing and adorning the more fruitful and humble plains, groves and forests, such as were never prophan'd by the inhumanity of edge-tools: Woods, whose original are as unknown as the Arcadians; like the goodly cedars of Libanus, *Psalm* 104, *arbores Dei*, according to the Hebrew, for something doubtless which they noted in the genius of those venerable places besides their meer bulk and stature: And verily, I cannot think to have well acquitted my self of this useful subject, till I shall have in some sort vindicated the honour of trees and woods, by shewing my reader of what estimation they were of old for their divine, as well as civil uses; at least refresh both him, and my self, with what occurs of historical and instructive amongst the learned concerning them. And first, standing woods and forests were not only the original

habitations of men, and for defence and fortresses, but the first occasion of that speech, polity and society which made them differ from beasts. This, the architect [1] Vitruvius ingeniously describes, where he tells us that the violent percussion of one tree against another forced by an impetuous wind, setting them on fire, the flame did not so much surprize and affright the salvage foresters, as the warmth, which (after a little gazing at the unusual accident) they found so comfortable: This (says he) invited them to approach it nearer, and as it spent and consum'd, by signs and barbarous tones (which in process of time were form'd into significant words) to encourage one another to supply it with fresh combustibles: By this accident the wild people, who before were afraid of one another, and dwelt asunder, began to find the benefit and sweetness of society, mutual assistance, and conversation, which they afterwards improv'd, by building houses with those trees, and dwelling nearer together: From these mean and imperfect beginnings they arriv'd in time to be authors of the most polish'd arts, establish'd laws, peopl'd nations, planted countries, and laid the foundation of all that order and magnificence which the succeeding ages have enjoy'd : No more then let us admire the enormous moles and bridges of Caligula across to Baiæ; or that of Trajan over the Danubius, stupendous work of stone and marble, to the adverse shores; whilst our timber and our trees making us bridges to the furthest Indies and Antipodes, land us into new worlds: In a word (and to speak a bold and noble truth) trees and woods have twice sav'd the whole world; first by the Ark, then by the Cross; making full amends for the evil fruit

[1] Vitruv. *l.* 2. *c.* I.

of the tree in Paradise, by that which was born on the tree in Golgotha. But that we may give an account of their sacred, and other uses of these venerable retirements, we will next proceed to describe what those places were.

2. Though *Silva* was the more general name, denoting a large tract of wood, or trees, the *inciduae* and *caeduae;* yet there were several other titles attributed to greater or lesser assemblies of them : *Domus Silvae* was a summer-house ; and such was Solomon's ᵒἶκος δρυμοῦ. 1 *Reg.* VII. 2. As when they planted them for pleasure and shade only, they had their *nemora;* and as we our parks, for the preservation of game, and particularly venison, &c. their *saltus,* and *silva invia,* secluded for the most part from the rest, &c. But among authors we meet with nothing more frequent, and indeed more celebrated, than those arboreous amenities and plantations of woods, which they call'd *luci;* and which, though sometimes we confess, were restrain'd to certain peculiar places, for devotion (which were never to be fell'd); yet were they also promiscuously both used, and taken for all that the wide forest comprehends, or can signifie. To dismiss a number of critics, the name *lucus* is deriv'd by Quintilian and others who delight to play with words (by *antiphrasis*) *a minime lucendo* because of its density, `

.... *nulli penetrabilis astro.* [1]

whence Apuleius us'd *luco sublucido;* and the poets, *sublustri umbra:* Others (on the contrary) have taken it for light in the masculine; *umbra non quia minime,*

[1] Vide Just. Lipsium in Germaniam Taciti prolixe satis.

sed quia maxime luceat; by so many lamps, suspended in them before the shrine; or because they kindled fires, by what accident unknown :

> [1] ....Whether it were
> By lightning sent from heaven, or else there
> The salvage-men in mutual wars and fight,
> Had set the trees on fire, their foes t'affright.

Or whether the trees set fire on themselves,

> [2] When clashing boughs thwarting, each other fret.

For such accidents, and even the very heat of the sun alone has kindled wonderful conflagrations: Or haply (and more probably) to consume their sacrifices, we will not much insist. The poets it seems, speaking of Juno, would give it quite another original, and tune it to their songs invoking Lucina, whilst the main and principal difference consisted not so much in the name, as the use and dedication, which was for silent, awful, and more solemn religion, (*silva, quasi silens locus*) to which purpose they were chiefly *manu consiti*, such as we have been treating of, entire, and never violated with the ax: Fabius calls them *sacros ex vetustate*, venerable for their age ; and certain it is, they had of very great antiquity been consecrated to holy uses, not only by superstitious persons to the Gentile deities and heroes, but to the true God, by the patriarchs themselves, who *ab initio* (as is presum'd) did frequently retire to such places to serve him in,

---

[1] .... Seu cæli fulmine misso,
Sive quod inter se bellum silvestre gerentes
Hostibus intulerant ignem, formidinis ergo, &c.
> *Lucret.* lib. v. 1243.

[2] Mutua dum inter se rami stirpesque teruntur.

compose their meditations, and celebrate sacred mysteries, prayers, and oblations; following the tradition of the Gomerites or descendants of Noah, who first peopl'd Galatia and other parts of the world after the universal [1] Deluge. from hence some presume that even the ancient Druids had their origin: But that Abraham might imitate what the most religious of that age had practis'd before him, may not be unlikely; for we read he soon planted himself and family at the *Quercetum* of Mambre, Gen. 13. where, as [2] Eusebius, *Eccl. Hist.* l. 1. c. 18, gives us the account, he spread his pavilions, erected an altar, offer'd and perform'd all the priestly rites; and there, to the immortal glory of the oak, or rather arboreous temple, he entertained God himself. Isidor, St Hierom,[3] and Sozomen report confidently, that one of the most eminent of those trees remained till the reign of the great Constantine, (and the stump till St Hierom) who founded a [4] venerable chappel under it; and that both the Christians, Jews, and Arabs, held a solemn anniversary or station there, and believed that from the very time of Noah, it had been a consecrated place: Sure we are, it was about some such assembly of trees, that God was pleas'd first of all to appear to the Father of the faithful, when he established the covenant with him, and more expresly, when removing thence (upon confirming the league with Abimelech, *Gen.* 21. and settling at Beersheba) he design'd an express place for God's divine service: For there, says the sacred text, he planted a

---

[1] See the learned Pezron *Antiq. fuse.*

[2] Euseb. lib. V. cap. 19. *Demonstr. Evang. ubi de Terebintho.* Hieronymus, *de locis Hebraicis,* &c.

[3] Hierom. in *Epitaph. Paul.* vide & Erasm. *Schol. in Ep. ad. Pamachium.*

[4] See the Emperor's Rescript to Bish. Macarius, &c. for the demolition of the idol worshipp'd there; and the building of a magnificent church. Euseb. *de vit. Constant.* lib. III. cap. 50.

grove, and called upon the name of the Lord. Such
another tuft we read of (for we must not always restrain
it to one single tree) when the patriarch came to Elon
Moreh, *ad convallem illustrium:* But whether that were
the same in which the high-priest reposited the famous
stone, after the exhortation mention'd, *Joshua* 24. 26,
we do not contend; under an oak says the Scripture,
and it grew near the Sanctuary, and probably might
be that which his grand-child consecrated with the
funeral of his beloved Rebecca, *Gen.* 35. For 'tis
apparent by the context, that there, God appeared to
him again: So Grotius upon the words (*subter quercum*)
illam ipsam (says he) *cujus mentio,* Gen. 35. 4, *in historia
Jacobi & Judae;* and adds, *is locus in honorem Jacobi diu
pro templo fuit.* That the very spot was long after
us'd for a temple in honour of him; and that place
which Sozomen calls Terebinthum, from certain trees
growing there as ancient as the world it self, says
Josephus *de bell. Jud.* l. 5. Others report that this
tree sprung from a staff, which one of the angels, who
appear'd to the patriarch, fixed in the ground: So Geor.
Syncellus *in Chronico. Mirum vero est* (says Valesius
on this passage of Eusebius) *cum quercus ibidem fuerit,
sub qua Abraham tabernaculum posuerit,* (*ut legitur in
cap.* 18. Gen.) *cur locus ipse a terebintho potius quam
a quercu nomen acceperit.* In the mean time, as to the
prohibition, XVI. *Deut.* 21, whether this patriarchal
devotion in groves, and under arboreous shades, was
approv'd by God, till there was a fixed altar, and his
ceremonial worship confin'd to the Tabernacle and
Temple, I think needs be no [1] question.

    3. If we therefore now would track the religious
esteem of trees and woods, yet farther in Holy Writ,

---

[1] D. Doughty. *Analecta Sacra, Excurs.* XIII.

we have that glorious vision of Moses in the fiery
thicket; and it is not to abuse or violate the text, that
Moncæus and others, interpret it to have been an intire
grove, and not a single bush only, which he saw as
burning, yet unconsum'd. *Puto ego* (says my author)
rubi vocabulo non quidem rubum aliquem unicum & solita-
rium significari, verum rubetum totum, aut potius fruticetum,
quomodo de quercu Mambre pro querceto toto docti intel-
ligunt. Now that they worshipped in that place soon
after their coming out of Ægypt, the following story
shews; and the Feast of Tabernacles had some re-
semblance of patriarchal devotion under trees, though
but in temporary groves and shades in manner of booths,
yet celebrated with all the refreshings of the forest;
and from the very infancy of the world in which
Adam was entertain'd in Paradise, and Abraham (as
we noted) receiv'd his divine guests, not in his tent,
but under a tree, an oak, (*triclinium angelicum,* the
angels dining-room) all intelligent persons have im-
brac'd the solace of shady arbours, and all devout
persons found how naturally they dispose our spirits
to religious contemplations: For this, as some conceive, they much affected to plant their trees in circles,
and gave that capacious form to the first temples,
observ'd not only of old,[1] but even at this day by the
Jews, as the most accommodate for their assemblies;
or, as others, because that figure most resembl'd the
universe, and the heavens: *Templum a templando,* says
a knowing critic; and another, *templum est nescio quid*
immane, atque amplum; such as Arnobius speaks of, that
had no roof but heaven, till that sumptuous fabric of
Solomon was confin'd to Jerusalem, and the goodliest
cedars, and most costly woods were carried thither to

[1] XXIII. *Levit.* 40.

form the columns, and lay the rafters; and then, and
not till then, was it so much as schism that I can find,
to retire to groves for their devotion, or even to
Bethel it self.

2. In such recesses were the ancient oratories and
proseuche, built theatre-wise, *sub dio*, at some distance
from the cities, xvi. *Acts ;* and made use of even
amongst the Gentiles, as well as the people of God ;
(nor is it always the less authentical for having been
the guise of nations) hence that of Philo, speaking of
one who πάσας Ἰουδαίων προσευχὰς ἐδενδροτόμησε, &c. had
fell'd all the trees about it; and such a place the
satyrist means, where he asks, *In qua te quaero proseucha?*
because it was the rendezvous also, where poor people
us'd to beg the alms of devout and charitable persons;[1]
so as it was esteem'd piacular for any to cut down so
much as a stick about them, unless it were to build
them, when with the Psalmist, men had honour according to their forwardness of repairing the houses
of God in the land, upon which account it was lawful
to lift up axes against the goodliest trees in the forest ;
but those zealous days are past ;

> [2] Now temples shut, and groves deserted lie,
> All gold adore, and neglect piety.

In the mean time, that which came nearest to the
Scaenopegia of the Jews, and other solemnities, call'd
by the Romans *Umbrae ;* as those in *Neptunalibus* are
describ'd by the poet,

---

[1] See Tirinus, our Mede, Ainsworth. *Diatrib.* on XXIV. *Josh.* 26. Valesius *Annos.* in lib. 2. *Hist. Eccles.* Euseb. p. 28.

[2] Et nunc desertis cessant sacraria lucis,
Aurum omnes victâ jam pietate colunt.
Propert.

[1] All sorts together flock; and on the ground
Display'd, each fellow with his mate drinks round.
Some sit in open air, some build their tents;
And some themselves in branchy arbors fence.

Plutarch speaking of the anniversary feast of
Bacchus, plainly resembles it to that of the Tabernacles, carrying about Θύρσους φοινικῶν, branches of palm,
citron, and other trees, as Josephus describes the
Jewish festival:[2] The custom (for ought I know) still
kept up in many places of our country, and abroad on
May-Day (and about the time of the year) when the
young men and maidens, like the pagan θυρσοφορία
go out into the woods and copp'ces, cut down and
spoil [3] young springers, to dress up their May-booth,
and dance about the pole, as in pictures we see the
wanton Israelites about the molten calf. For thus,
as we noted, those rites commanded by God, came to
be prophaned, and the retireness of groves and shades
for their opacousness, abus'd to abominable purposes,
and works of darkness: But what good, or indifferent
thing has not been subject to perversion? It is said
in the end of Isaiah, *Exprobratur Hebraeis quod in
opisthonais idolorum horti essent in quorum medio februa-
bantur;*[4] but how this is applicable to groves, does not
appear so fully; though we find them interdicted,
Deut. 16. 21. *Judg.* 6. 26. 2 *Chron.* 31. 3, &c. and
forbidden to be planted near the Temple.[5] And an
impure grove on Mount Libanus, dedicated to Venus,

[1] Plebs venit, ac virides passim disjecta per herbas
Potat & accumbit cum pare quisque sua:
Sub Jove pars durat, pauci tentoria ponunt;
Sunt quibus e ramis frondea facta casa est.
Ovid, Fast. lib. 3. (march.)
[2] *Sympos.* l. 2 q. 8.
[3] See cap. VIII. lib. III. sect. 5.
[4] Vide Seldenum *de jure nat. & gent. Heb.* l. 2. c. 6.
[5] Lil. & Gre. Gyraldum *de diis gent. Syntag.* 17.

was by an imperial edict of Constantine, extirpated;
but from the abuse of the thing to the non-use, the
consequence is not always valid, and we may note as
to this very particular, that where in divers places of
Holy Writ, the denunciation against groves is so
express, it is frequently to be taken but catachrestically,
from the wooden image or statue call'd by that name,
as our learned Selden makes out by sundry instances
in his *Syntagma de diis Syris*. Indeed the use of groves
upon account of devotion, was so ancient, and seem'd
so universal, that they consecrated not only real and
natural groves, but *lucos pictos*, artificial boscage and
representations of them.

The sum of all is, Paradise it self was but a kind
of nemorous temple, or sacred grove, planted by God
himself, and given to man, *tanquam primo sacerdoti*, the
word is one which properly signifies to serve or admin-
ister, *res divinas*, a place consecrated for sober discipline, and to contemplate those mysterious and
sacramental trees which they were not to touch with
their hands; and in memory of them, I am inclin'd
to believe, holy men (as we have shew'd in Abraham
and others) might plant and cultivate groves, where
they traditionally invok'd the Deity; and St. Hierom,
Chrysostom, Cyprian, Augustine, and other Fathers
of the Church greatly magnified these pious advantages; and Cajetan tells us, that from Isaac to Jacob,
and their descendants, they followed Abraham in this
custom: Solomon was a greater planter of them, and
had an house of pleasure or lodge in one of them for
recess: In such places were the monuments of their
saints, and the bones of their heroes deposited; for

---

[1] Vide Sanctium, Piscat. Grotium.

which David celebrated the humanity of the Gileadites, *in nemora Jabes*, as most sacred and inviolable. In such a place did the angel appear to Gideon; and in others, princes were inaugurated; so Abimelech, *Judic*.9. And the rabbins add a reason why they were reputed so venerable; because more remote from men and company, more apt to compose the soul, and fit it for divine actions, and sometimes apparitions, for which the first enclosures, and *sacra septa* were attributed to groves, [1] mountains, fountains of water, and the like solemn objects; as of peculiar sanctity, and as the old sense of all words denoting sanctity did import separateness, and uncommon propriety : See our learned Mede. For though since the Devil's intrusion into Paradise, even the most holy and devoted places were not free from his temptations and ugly stratagems ; yet we find our Blessed Saviour did frequently retire into the wilderness, as Elijah and St. John Baptist did before him, and divers other holy men ; particularly, the Θεωρητικοὶ, whom Philo [2] mentions ; a certain religious sect, who addicting themselves to contemplation, chose the solitary recesses of groves and woods, as of old the Rechabites, Essenes, primitive monks, (and other institutions) retired amongst the Thebaid desarts: And perhaps the air of such retired places may be assistant and influential, for the inciting of penitential expressions and affections ; especially where one may have the additional assistances of solitary grotts, murmuring streams, and desolate prospects. I remember that under a tree was the place of that admirable St. Augustine's solemn conversion, after all his importunate reluctances: I have often thought of it, and it is

[1] καὶ πρῶτον ὕλας ἀπενέμοντο, καὶ ὄρη ἀνέθεσαν. Lucian *de Sacrif*.
[2] Philo. lib. περὶ βίου θέας.

a melting passage, as himself has recorded it, *Con.* l. 8. c. 8. and he gives the reason, *solitudo enim mihi ad negotium flendi aptior suggerebatur.* And that indeed such opportunities were successful for recollection, and to the very reformation of some ingenuous spirits, from secular engagements to excellent and mortifying purposes, we may find in that wonderful relation of Pontianus's two friends, great courtiers of the time, as the same holy father relates it, previous to his own conversion. And here I cannot omit an observation of the learned Dr. Plot, in his (often-cited) *Nat. Hist. of Oxfordshire;* taking notice of two eminent religious houses, whose foundations were occasion'd by trees : The first, Oseney-Abby : The second, by reason of a certain tree standing in the meadows (where after was built the abby) to which a company of pyes were wont to repair, as oft as Editha the wife of Robert d'Oyly, came to walk that way to solace her self; for the clamorous birds did so affect her, that consulting with one Radulphus (canon of St. Fridiswid) what it might signifie, the subtle man advis'd her to build a monastery where that tree stood, as if so directed by the pyes in a miraculous manner : Nor was it long e'er the lady procur'd her husband to do it, and to make Radulphus (her confessor) first prior of it.

Such another foundation was caus'd by a tripple elm, having three trunks issuing from one root: Near such a tree as this was Sir Thomas White, Lord Mayor of London, warn'd by dream to erect a college for the education of youth, which he did ; namely, St. John's in Oxford, which with the very tree, still flourishes in that famous University. But of these enough, and perhaps too much.

6. We shall now in the next place endeavour to shew

how this innocent veneration to groves passed from the people of God to the Gentiles, and by what degrees it degenerated into dangerous superstitions : For the Devil was always God's ape, and did so ply his groves, altars, and sacrifices, and almost all other rites belonging to his worship, that every green tree was full of his abominations, and places devoted to his impure service;[1] *Hi fuere* (says Pliny, speaking of groves) *quondam numinum templa*, &c. ' These were of old ' the temples of the gods, and after that simple (but ' ancient custom) men at this day consecrate the ' fairest and goodliest trees to some deity or other; ' nor do we more adore our glittering shrines of gold ' and ivory, than the groves, in which with a profound and awful silence, we worship them.[2] Quintilian speaking of the veneration paid an old umbragious oak, adds, *In quibus grandia & antiqua robora jam non tantum habeat speciem, quantum religionem :* For in truth, the very tree it self was sometimes deified, and that Celtic statue of Jupiter no better than a prodigious tall oak, whence 'tis said the Chaldean theologues deriv'd their superstition towards it;[3] and the Persians we read, us'd that tree in all their mysterious rites. And as for wood in general, they paid it that veneration, for its maintaining their deity, (represented by their perennial fire) that they would not suffer any sort of wood to be us'd for coffins to inclose the dead in, (but in plates of iron) counting it a profanation. In short, so were people given up to this devilish and unnatural blindness, as to the offering of human sacrifices not to the tree-gods only, but to the trees themselves as real gods.

[1] Cyril. Alexand. in *Hos.* 4. 13; *Deut.* 16. 4; 2 *Reg.* 16. 4.
[2] Melchior Adamus *Hist. Eccles. de Sueconibus*, c. 234.
[3] Mariana in 2. *Paralip.* 28. 4.

[1] Each tree besprinkled is with human gore.

Procopius tells us plainly that the Sclavii worshipped trees and whole forests of them : See Jo. Dubravius, l. 1. *Hist. Bohem.* and that formerly the Gandenses did the like ; (see Surius the Legendary, 6. Feb. reports in the *Life of St. Amadus* :) So did the Vandals, says Albert Crantz ; and even those of Peru, as I learn from Acosta, l. 5. c. 11. But one of the first idols which procur'd particular veneration in them, was the Sidonian Ashteroth, who took her name *a Lucis*, as the Jupiter ἐνδενδρος amongst the Rhodians, the *Nemorensis Diana* or *Arduenna*, a celebrated deity, of this our island, for her patronage of wood and game,

Diva potens nemorum, terror silvestribus apris, &c.

as Gildas an ancient bard of ours has it ; so soon had men it seems degenerated into this irrational and stupid devotion, that Arch-fanatic Satan (who began his pranks in a tree) debauching the contemplative use of groves, and other solitudes. Nor were the heathens alone in this crime ; the Basilidians, and other hereticks, even amongst the Christians, did consecrate to the woods and the trees, their serpent-footed and barbarous ΑΒΟΡΑΞΑΣ, as it is yet to be seen in some of their mysterious talismans and *periapta* which they carried about.

But the Roman madness (like that which the prophet[2] derides in the Jews) was well perstring'd by Sedulius and others, for imploring these stocks to be propitious to them, as we learn in Cato *de R. R.* c. 113. 134, &c. Nor was it long after, (when they

[1] Omnisque humanis lustrata cruoribus arbos. Lucan, l. 3. 405.
[2] In opere Paschali.

were generally consecrated by Faunus) that they boldly set up his oracles and responses in these nemorous places : Hence the heathen chappels had the name of *fana*, and from their wild and extravagant religion, the professors of it fanatics ; a name well becoming some of our late enthusiasts amongst us ; who, when their quaking fits possess them, resemble the giddy motion of trees, whose heads are agitated with every wind of doctrine.

7. Here we may not omit what learned men have observ'd concerning the custom of prophets and persons inspir'd of old, to sleep upon the boughs and branches of trees; I do not mean on the tops of them, (as the salvages somewhere do in the Indies for fear of wild beasts in the night-time) but on matrasses and beds made of their leaves, *ad consulendum*, to ask advice of God. Naturalists tell us, that the *laurus*, and *agnus castus* were trees which greatly compos'd the fancy, and did facilitate true visions ; and that the first was specially efficacious πρὸς τοὺς ἐνθουσιασμούς, (as my author expresses it) to inspire a poetical fury:[1] Such a tradition there goes of Rebekah the wife of Isaac, in imitation of her father-in-law : The instance is recited out of an ancient ecclesiastical history by Abulensis;[2] and (what I drive at) that from hence the Delphic *Tripos*, the Dodonæan oracle in Epirus, and others of that nature had their originals: At this decubation upon boughs the satyrist seems to hint, where he introduces the gypsies :

[3] .... With fear
A cheating Jewess whispers in her ear,

[1] See Fulgent. *Mythol.* cap. 13. & Munsherum in *Comment.*
[2] See Hier. in *Trad. Heb.* 3 *Reg.* c. 4.
[3] Arcanam Iudæa tremens mendicat in aurem,
Interpres legum Solymarum, & magna sacerdos
Arboris, ac summi fida internuntia caeli.   *Juv. Sat. 6.*

And begs an alms : An high-priest's daughter she,
Vers'd in their Talmud, and Divinity ;
And prophesies beneath a shady tree.

Dryden.

For indeed the Delphic oracle (as Diodorus *l.* 16.
tells us) was first made *e lauri ramis*, of the branches
of laurel transferr'd from Thessaly, bended, and arched
over in form of a bower or summerhouse, a very
simple fabrick you may be sure : And Cardan I remember in his book *de Fato*, insists very much on
the dreams of trees for portents and presages, and
that the use of some of them do dispose men to visions.

8. From hence then began temples to be erected
and sought to in such places;[1] nay we find [2] sanction
for it among the laws of the XII tables : So as there
was hardly a grove without its temple, so had every
temple almost a grove belonging to it, where they
plac'd idols, altars and lights, endowed with fair re-
venues, which the devotion of superstitious persons
continually augmented : Such were those [3] *arbores
obumbratrices*, mention'd by Tertullian (*Apol.* cap. IX.)
on which they suspended their Ἀναϑήματα and devoted
things : And I remember to have seen something
very like this in Italy, and other parts, namely, where
the images of the B. Virgin, and other saints, have
been enshrin'd in hollow and umbragious trees, frequented with much veneration ; which puts me in
mind of what that great traveller Pietro della Valle
relates, where he speaks of an extraordinary cypress,
yet extant, near the tomb of Cyrus, to which at this

[1] Vide Annium Viterb. *l.* 17. *fol.* 158.
[2] Cic. *de lege. l.* 2.
[3] See Aristophanes, *Schol. ad Pluti verba :* καὶ ταῦτα πρὸς τὸ μέτωπον, &c.
........ὅτι ἐπὶ τῶν κοτίνων καὶ ἄλλων δένδρων πανταχοῦ ἐν τοῖς ἱεροῖς προσπατταλεύουσι
τὰ ἀναθήματα. To which add, Apul. Miles. VI. Videt dona speciosa, & lacinias
auro literatas, ramis arborum postibusque suffixas.

day many pilgrimages are made, and speaks of a gummy transudation which it yields, that the Turks affirm to turn every Friday into drops of blood : The tree is hollow within, adorn'd with many lamps, and fitted for an oratory ; and indeed some would derive the name *lucus* a grove, as more particularly to signifie such enormous and cavernous trees, *quod ibi lumina accenderentur religionis causa:* But our author adds, the Ethnics do still repute all great trees to be divine, and the habitation of souls departed : These the Persians call *Pir* and *Imàm.* Perhaps such a hollow tree was that asylum of our poet's hero, when he fled from his burning Troy.

[1] .... an ancient cypress near,
Kept by religious parents many a year.

For that they were places of protection, and privileg'd like churches, and altars, appears out of Livy, and other good authority : Thus where they introduce Romulus encouraging his new colony,

[2] So soon as e're the grove he had immur'd
Haste hither (says he) here you are secur'd.

Such a sanctuary was the Aricina,[3] and suburban Diana, call'd the *nemorale templum,* and divers more which we shall reckon up anon. Lucian in his *dea Syri* speaks of these temples and dedications in their groves among the Egyptians : *Lucus in urbe fuit, &c.* and what follows ? *Hic templum—* and since they could

[1] .... juxtaque ; antiqua cupressus
Religione patrum multos servata per annos.
Æn. 2.
[2] ....Ut saxo lucum circumdedit alto
Quilibet, huc, dicit, confuge, tutus eris.
[3] Virg. 6. *Eclog.* and 1 *Æneid.* vide Fab. l. 3. Semest. c. 1.

not translate the grove with the idol, they [1] carv'd
out something like it, which the superstitious people
bought, carried home, and made use of representing
those venerable places, in which they had the images
of some feign'd deity (suppose it Tellus, Baal or
Priapus) ; and such was the Jupiter ἔνδενδρος of the
Rhodians, Bacchus of the Bœotians, the Sidonian
Ashteroth : And the women mentioned 2 *Reg.* 23. 7.
who are said to weave hangings and curtains for the
grove, were no other than makers of tentories, to
spread from tree to tree, for the more opportune and
secret perpetration of those impure rites and mysteries,
which (without these coverings) even the opacousness
of the places were not obscure enough to conceal.

9. The famous Druids, or [2] Saronides, whom the
learned Bochart from Diodorus, proves to be the
same, derived their oak-theology, namely, from that
spreading and gloomy shading tree, probably the grove
at Mamre, XIII. *Gen.* 18.   How their mysteries were
celebrated in their woods and forests, is at large to
be found in Caesar, Pliny, Strabo, Diodorus, Mela,
Apuleius, Ammianus, Lucan, Aventinus, and innumerable other writers, where you will see that they
chose the woods and the groves, not only for all their
religious exercises, but their courts of justice ; as the
whole institution and discipline is recorded by Caesar,
l. 6. and as he it seems found it in our country of
Britain, from whence it was afterwards translated
into Gallia : For he attributes the first rise of it to
this once happy island of groves and oaks ; and affirms,

---

[1] Luci dicuntur, non modo collectio arborum, &c. sed etiam sciagraphiae sive
delineationes lucorum in tabella : See the Annotation on *Isa.* 17. collated with
2 *Reg.* 23. 6. *Crit. Sacr.* for they brought the grove out of the temple, and burnt
it, which clearly shews it was the picture or image of the grove, and not the trees
themselves.

[2] Canc. l. 1. cap. 42. Selden *Jani Angl. fac.* cap. 2.

that the ancient Gauls travelled hither for their initiation. To this Tacitus assents, 14 *Annal.* and our most learned critics vindicate it both from the Greeks and French, impertinently challenging it : But the very name it self, which is purely Celtic, does best decide the controversie : For though δρῦς be *quercus* ; yet Vossius skilfully proves that the Druids were altogether strangers to the Greeks; but what comes yet nearer to us, *dru, fides* (as one observes) begetting our now antiquated trou, or true, makes our title the stronger : Add to this, that amongst the Germans it signified no less than God himself ; and we find *drutin*, or *trudin* to import divine, or faithful in the Othfridian gospel, both of them sacerdotal expressions. But that in this island of ours, men should be so extreamly devoted to trees, and especially to the oak, the strength and defence of all our enjoyments, inviron'd as we are by the seas, and martial neighbours, is less to be wonder'd,

[1] Our Brittish Druids not with vain intent,
Or without Providence did the oak frequent ;
That Albion did that tree so much advance
Nor superstition was, nor Ignorance,
Those priests divining even then, bespoke
The mighty triumphs of the royal oak :
When the sea's empire with like boundless fame,
Victorious Charles the son of Charles shall claim.

as we find the prediction gloriously followed by our

[1] Non igitur Dryadæ nostrates pectore vano,
Nec sine consulto coluerunt numine quercum ;
Non illam Albionis jam tum celebravit honore
Stulta superstitio, venturive inscia secli
Angliaci ingentes puto prævidisse triumphos
Roboris, imperiumque maris quod maximus olim
Carolides vastâ victor ditione teneret.

Coulei l. 6. pl.

ingenious poet, where his Dryad consigns that sacred
depositum to this monarch of the forest, the oak;
than which nothing can be more sublime and rap-
turous, whilst we must never forget that wonderful
Providence which saved this forlorn and persecuted
Prince, after his defeat at Worcester, under the shelter
of this auspicious and hospitable tree alone ; When

> .... All the countries fill'd
> With enemies troops, in every house and grove,
> His sacred head is at a value held,
> They seek, and near, now very near they move.
>
> What should they do ? They from the danger take
> Rash, hasty counsel ; yet from heav'n inspir'd,
> A spacious oak he did his palace make,
> And safely in its hollow womb retir'd.
>
> The loyal tree its willing boughs inclin'd
> Well to receive the climbing royal guest,
> (In trees more pity than in men we find)
> And its thick leaves into an arbor prest.
>
> A rugged seat of wood became a throne,
> The obsequious boughs his canopy of state :
> With bowing tops the tree their King did own,
> And silently ador'd him as he sate.

But to return to the superstition we were speaking
of (since utterly abolish'd) till the reign of Claudius,
as appears by Suetonius ; yet by Tacitus they continued here in Britain under Nero, and in Gaule till
Vitellius, as is found by St. Gregory writing to Q.
Brunehant, about the prohibiting the sacrifices and
worship which they paid to trees : Which Sir John
Ware affirms continued in Ireland till Christianity
came in.

10. From those silvan philosophers and divines (not to speak much of the Indian Brahmans, or ancient gymnosophists) 'tis believed that the great Pythagoras might institute his silent monastery ; and we read that Plato entertained his auditors amongst his walks of trees, which were afterward defac'd by the inhumanity of Sylla, when as Appian tells us, he cut down those venerable shades to build forts against Piræus : And another we find he had, planted near Anicerides with his own hands, wherein grew that celebrated *platanus* under which he introduces his master Socrates discoursing with Phædrus *de pulchro* : Such another place was the Athenian Cephisia, as Agellius describes it : We have already mention'd the stately *xysta*, with their shades, in cap. 23. Democritus also taught in a grove, as we find in that of Hippocrates to Damagetus, where there is a particular tree design'd *ad otium literarum;* and I remember Tertullian calls these places *Studia opaca :* Under such shades and walks was at first the famous Academia, esteem'd so venerable, as it was by the old philosopher, prophane so much as to laugh in it, see Laertius, Ælian, &c. I could here tell you of Palæmon, Timon, Apollonius, Theophrastus, and many more that erected their schools in such colleges of trees, but I spare my reader ; I shall only note, that 'tis reported of Thucydides, that he compiled his noble History in the Scaptan Groves, as Pliny writes ; and in that matchless piece *de Oratore,* we shall find the interlocutors to be often under the *platanus* in his Tusculan villa, where invited by the freshness and sweetness of the place, *admonuit* (says one of them) *me haec tua* platanus *quae non minus ad opacandum hunc locum patulis est diffusa ramis, quam illa, cujus umbram secutus est* Socrates, *quae mihi videtur non*

cc

tam ipsa aquula, quae describitur, quam Platonis *oratione crevisse &c.* as the orator brings it in, in the person of one of that meeting.

I confess Quintilian seems much to question whether such places do not rather perturb and distract from an orator's [1] recollection, and the depths of contemplation: *Non tamen* (says he) *protinus audiendi, qui credunt aptissima in hoc nemora, silvasque, quod illa caeli libertas, locorumque amoenitas, sublimem animum, & beatiorem spiritum parent : Mihi certe jucundus hic magis, quam studiorum hortator videtur esse secessus : Namque illa ipsa quae delectant, necesse est avocent ab intentione operis destinati:* He proceeds........ *Quare silvarum amoenitas, & praeterlabentia flumina, & inspirantes ramis arborum aurae, volucrumque cantus & ipsa late circumspiciendi libertas, ad se trahunt ; ut mihi remittere potius voluptas ista videatur cogitationem quam intendere.* But this is only his singular suffrage, which as conscious of his error, we soon hear him retract, when he is by and by as loud in its praises, as the places in the world the best fitted for the diviner rhetoric of poetry : But let us admit another [2] to cast in his symbol for groves : *Nemora* (says he) *& luci, et secretum ipsum, tantum mihi afferunt voluptatem, ut inter praecipuos carminum fructus numerem, quod nec in strepitu, nec sedente ante ostium litigatore, nec inter sordes & lacrimas reorum componuntur : Sed secedit animus loca pura, atque innocentia, fruiturque sedibus sacris.*

Whether this were the effect of the incomparable younger Pliny's Epistle [3] to this noble historian, I know not; but to shew him by his own example how study and forest-sport may consist together, he tells him

---

[1] See this most elegantly discuss'd in a Greek Epistle of Budæus to his brother, ep. I.

[2] Tacitus.

[3] Plin. Ep. VI. Lib. I. Cornelio Tacito.

how little the noise of the chasers and bawling dogs
disturbed him, when at any time he indulged himself
that healthful diversion : ' So far was he from being
' idle, and losing time, that beside his javelin and
' hunting-pole, he never omitted to carry his style
' and table-book with him, that upon any intermission,
' whilst he now and then sate by the toil and nets, he
' might be ready to note down any noble thought,
' which might otherwise escape him : The very
' motions (says he) and agitation of the body in the
' wood and solitude, *magna cogitationis incitamenta*
' *sunt :* I know, my friend (says he) you'll smile at it,
' however take my counsel ; be sure never to carry
' your bottle and bisque into the field, without your
' *pugillares* and tablet ; you'll find as well Minerva as
' Diana in the woods and mountains.

And indeed the Poets thought of no other heaven
upon earth, or elsewhere ; for when Anchises was
setting forth the felicity of the other life to his son,
the most lively description he could make of it was
to tell him,

[1]............ We dwell in shady groves.

and that when Æneas had travelled far to find those
happy abodes,

[2] They came to groves, of happy souls the Rest,
To ever-greens, the dwellings of the blest.

Such a prospect he gives us of his Elisium ; and
therefore wise and great persons had always these

---

[1] .... Lucis habitamus opacis.
[2] Devenere locos laetos, & amoena vireta
Fortunatorum nemorum, sedesque beatas.

sweet opportunities of recess, their *domos silvae*, as we read, 2 *Reg.* 7. 2. which were thence called Houses of Royal Refreshment, or as the Septuagint, οἴκους δρυμοῦ, not much unlike the lodges in divers of our noblemens parks and forest-walks; which minds me of his choice in another poem,

> [1] In lofty towers let Pallas take her rest,
> Whilst shady groves 'bove all things please us best.

And for the same reason Maecenas

> [2] .... Chose the broad oak........

And as Horace bespeaks them,

> [3] Me the cool woods above the rest advance
> Where the rough satyrs with the light nymphs dance.

And Virgil again,

> [4] Our sweet Thalia loves, nor does she scorn
> To haunt umbragious groves.....................

Or as thus expressed by Petrarch,

> [5] .... The Muse her self enjoys
> Best in the woods, verse flies the city noise.

So true is that of yet as noble a poet of our own :

> .... [1] Pallas quas condidit arces,
> Ipsa colat, nobis placeant ante omnia silvae.
>
> Eclog. 2.
>
> [2] Maluit umbrosam quercum ....
>
> [3] .... Me gelidum nemus
> Nympharumque leves cum satyris chori,
> Secernunt populo ....
>
> [4] Nostra nec erubuit silvas habitare Thalia.
>
> [5] Silva placet musis, urbs est inimica poetis.

As well might corn, as verse in cities grow,
In vain the thankless glebe we plow and sow,
Against the unnatural soil in vain we strive,
'Tis not a ground in which these plants will thrive.

Cowley.

When it seems they will bear nothing but nettles and
thorns of satyrs, and as Juvenal says, by indignation
too ; and therefore ' almost all the poets, except those
' who were not able to eat bread without the bounty
' of great men ; that is, without what they could get
' by flattering them (which was Homer's and Pindar's
' case) have not only withdrawn themselves from the
' vices and vanities of the great world, into the inno-
' cent felicities of gardens, and groves, and retiredness,
' but have also commended and adorned nothing so
much in their never-dying Poems. [1] Here then is the
true Parnassus, Castalia, and the Muses, and at every
call in a grove of venerable oaks, methinks I hear the
answer of an hundred old Druids, and the bards of
our inspired ancestors.

In a word, so charm'd were poets with those natural
shades, especially that of the *platanus*, that they
honour'd temples with the names of [2] groves, though
they had not a tree about them : Nay sometimes,
one stately tree alone was so rever'd : And of such a
one there is mention in a fragment of an inscription
in a garden at Rome, where there was a temple built
under a spreading beech-tree, sacred to Jupiter, under
the name of *Fagutalis*.

Innumerable are the testimonies I might produce
in behalf of groves and woods out of the poets,

---

[1] Juvenal *Sat*. VII. Pers. *Sat*. I.
[2] ἄλση καλοῦντες τὰ ἱερὰ πάντα κἂν ᾖ ψιλὰ οἱ ποιηταὶ κοσμοῦσιν. Strab. *l*. 9.

Virgil, Gratius, Ovid, Horace, Claudian, Statius,
Silius, and others of later times, especially the divine
Petrarch : (for *Scriptorum chorus omnis amat nemus*)
were I minded to swell this charming subject, beyond
the limits of a chapter : I think only to take notice,
that theatrical representations, such as were those of
the Ionian call'd *Andria* ; the scenes of Pastorals,
and the like innocent rural entertainments, were of
old adorned and trimm'd up *e ramis & frondibus, cum
racemis & corymbis*, and frequently represented in
groves, as the learned Scaliger shews : And here the
most beloved and coy mistress of [1] Apollo rooted; and
the noblest raptures have been conceiv'd in the walks
and [2] shades of trees, and poets have composed verses
which have animated men to heroic and glorious
actions; here orators (as we shewed) have made their
panegyrics, historians grave relations, and the profound
philosophers loved here to pass their lives in repose
and contemplation ; and the frugal repasts..........
mollesque sub arbore somni, were the natural and chast
delights of our fore-fathers, so sweetly describ'd by
Papinius,

> Subter opaca quies vacuusque silentia servat
> Horror, & exclusæ pallet mala lucis imago
> Nec caret umbra Deo......
> Arboribus suus horror inest, quin ipse sacerdos
> Accessus Dominumque timet deprehendere luci.

12. Nor were groves thus only frequented by the
great scholars, and the great wits, but by the greatest
statesmen and politicians also : Thence that of Cicero
speaking of Plato, with Clinias and Megillus, who

---

[1] *Poetice*, lib. I. cap. 21.
[2] See Wower. *de Umbra*, cap. 26. Bisciola *Horae subcis.* cap. 9.

were us'd to discourse *de rerumpublicarum institutis, &* *optimis legibus*, in the groves of cypress, and other umbrageous recesses: It was under a vast oak growing in the park at St. Vincent's, near Paris, that St. Louis was us'd to hear complaints, determine causes, and do justice to such as resorted thither : And we read of solemn treaties of peace held under a flourishing elm between Gisors and Trier, which was afterwards fell'd by the French King Philip in a rage against King Henry II. not agreeing to it. Nay they have sometimes been known to crown their kings under a goodly tree, or some venerable grove where they had their stations and conventions ; for so they chose Abimelech, see Tostatus upon *Judges* 9. 6. and I read (in *Chronicon* Jo. Bromton) that Augustine the monk (sent hither from the Pope) held a kind of council under a certain oak in the West of England, and that concerning the great question, namely the right celebration of Easter, and the state of the Anglican-Church, *&c.* where also 't is reported he did a great miracle. In the mean time I meet with but one instance where this goodly tree has been (in our country) abus'd to cover impious designs, as was that of the arch-rebel Kett, who in the reign of King Edw. VI. (becoming leader to that fanatick insurrection in Norfolk), made an oak (under the specious name of [1] reformation), the court, counsel-house, and place of convention, whence he sent forth his trayterous edicts : The history and event of which, to the destruction of the rebel and his followers, together with the sermon, (call it speech or what you please) which our then young Matth. Parker, (afterward the venerable and learned archbishop of

[1] *Quercus Reformationis.*

Canterbury) boldly pronounced on it, to reduce them
to obedience, is most elegantly described in Latin,
and in a style little inferior to the ancients, by our
country-man Alexander Nevyll, in his *Kettus, sive de
furoribus Norfolciensium Ketto duce.* [1] But to return ;
the Athenians were wont to consult of their gravest
matters, and publick concernments in groves : Famous
for these assemblies were the Ceraunian, and at Rome
the Lucus Petilinus, the Farentinus, and others, in
which there was held that renowned Parliament after
the defeat of the Gauls by M. Popilius : For it was
supposed that in places so sacred, they would faithfully and religiously observe what was concluded
amongst them.

> In such green palaces the first kings reign'd,
> Slept in their shades, and angels entertain'd :
> With such old counsellors they did advise,
> And by frequenting sacred groves, grew wise ;
> Free from th' impediments of light and noise,
> Man thus retir'd, his nobler thoughts employs.
>
> *Mr. Waller.*

As our excellent poet has described it : And amongst
other weighty matters, they treated of matches for
their children, and the young people made love in
the cooler shades, and ingraved their mistresses names
upon the bark, [2] *tituli aereis literis insculpti,* as Pliny
speaks of that ancient Vatican *ilex,* and Euripides in
Hippolyto, where he shews us how they made the
incision, whisper their soft complaints like that of
Aristænetus, Τοῖα δὲ εἶϑε ὦ δένδρα, &c. and wish that it
had but a soul and voice to tell Cydippe, the fair
Cydippe, [3] how she was beloved : And doubtless this

---

[1] Edit. 8vo. Lond. 1582.
[2] L. 16. E 44. Arist. l. *Ep.* 10.
[3] Vide Symmach. l. 4. *Ep.* 28.

character was ancienter than that in paper ; let us
hear the amorous poet leaving his young couple thus
courting each other,

> [1] My name on bark engraven by your fair hand,
> Oenone, there, cut by your knife does stand ;
> And with the stock my name alike does grow,
> Be't so, and my advancing honour show.

which doubtless he learnt of Maro describing the
unfortunate Gallus.

> [2] There on the tender bark to carve my love;
> And as they grow, so may my hopes improve.

and these pretty monuments of courtship I find were
much used on the cherry-tree (the wild one, I suppose)
which has a very smooth rind, as the witty Calphurn-
ius,

> [3] Repeat, thy words on cherry-bark I'll take,
> And that red skin my table-book will make.

Let us add the sweet [4] Propertius,

> Ah quoties teneras resonant mea verba sub umbras,
> Scribitur & vestris Cynthia corticibus.

And so deep were the incisions made, as that of [5]
Helena on the platan (ὡς πα ρίων πς Ανγνοίη) that one

---

[1] Incisæ servant a te mea nomina fagi,
   Et legor, Oenone, falce notata tua :
  Et quantum trunci, tantum mea nomina crescunt:
    Crescite, & in titulos surgite rite meos.   *Ovid, Ep.* 5.

[2] .... tenerisque meos incidere amores
  Arboribus : Crescent illæ, crescetis amores.   *Eclog.* 10.

[3] Dic age, nam cerasi tua cortice verba notabo.
  Et decisa feram rutilanti carmina libro.

[4] Lib. I, *Elegia* XVIII.
[5] Theocrit. *Epithal. Helenae*, Idyll. 18.

might run and read them. And thus forsaken lovers
appeal to pines, beeches, and other trees of the forest :
But we have dwelt too long on these trifles ; omitting
also what we might relate of feasting, banqueting,
and other splendid entertainments under trees, nay
sometimes in the very bodies of them : But we will
now change the scene as the Ægyptians did the mirth
of their guests, when they served in a scull to make
them more serious.   For, thus

13. Amongst other uses of groves, I read that some
nations were wont to hang, not malefactors only, but
their departed friends, and those whom they most
esteemed, upon trees, as so much nearer to heaven,
and dedicated to God ; believing it far more honourable than to be buried in the earth ; and that some
affected to repose rather in these woody places,
Propertius seems to bespeak,

> [1] The Gods forbid my bones in the high road
> Should lie, by every wandring vulgar trod ;
> Thus buried lovers are to scorn expos'd,
> My tomb in some by-arbor be inclos'd.

The same is affirmed of other Septentrional people
by Chr. Cilicus *de Bello Dithmarsico*, l. 1.   It was
upon the trunk of a knotty and sturdy oak, the
ancient heroes were wont to hang the arms and
weapons taken from the enemy, as trophies, as appears
in the yet remaining stump of Marius at Rome, and
the reverses of several medals.   Famous for this,
was the pregnant oleaster which grew in the *forum*
of Megara, on which the heroes of old left their

---

[1] Dî faciant mea ne terrâ locet ossa frequenti
    Quâ facit assiduo tramite vulgus iter ;
   Post mortem tumuli sic infamantur amantum,
    Me tegat arboreâ devia terra comâ.

shields and bucklers, and other warlike harness, 'till
in process of time, it had cover'd them with successive coats of bark and timber, as it was afterwards
found, when Pericles sack'd the city; which the
oracle predicted should be impregnable, 'till a tree
should bring forth [1] armour. We have already
mention'd Rebekah, and read of kings themselves
that honoured such places with their sepulchres :
What else should be the meaning of 1 *Chro.* 10. 12.
when the valiant men of Jabesh interr'd the bones of
Saul and Jonathan under the oak ? Famous was the
Hymethian cœmeterie where Daiphron lay: Ariadne's
tomb was in the Amathusian grove in Crete, now
Candie ; for they believed that the spirits and ghosts
of men delighted to expatiate, and appear in such
solemn places, as the learned Grotius notes from
Theophylact, speaking of the daemons, upon *Mat.* 8.
20; for which cause Plato gave permission, that trees
might be planted over graves, to obumbrate and
refresh them: The most ancient *conditoria* and burying-
places, were in such nemorous solitudes : The Hypo-
gaeum in Macpela, purchas'd by the patriarch
Abraham of the sons of Heth, *Gen.* XXIII. for Sarah,
his own dormitory, and family's sepulchre ; was
convey'd to him, with particular mention, ver. 3. of
all the trees and groves about it ; and the very first
precedent I ever read, of conveying a purchase by a
formal deed.

Our Blessed Saviour, (as we shall shew) chose the
garden sometimes for his oratory, and dying, for the
place of his sepulchre ; and we do avouch for many
weighty causes, that there are none more fit to bury
our dead in, than in our gardens and groves, or airy

[1] Diodor. Sic. lib. 12.

fields, *sub dio;* where our beds may be decked and carpeted with verdant and fragrant flowers, trees, and perennial plants, the most natural and instructive hieroglyphics of our expected resurrection and immortality ; besides what they might conduce to the meditation of the living, and the taking off our cogitations from dwelling too intently upon more vain and sensual objects ; that custom of burying in churches, and near about them (especially in great and populous cities) being both a novel presumption, undecent, sordid, and very prejudicial to health ; and for which I am sorry 'tis become so customary. Graves and sepulchres were of old made and erected by the sides of the most frequented high-ways, which being many of them magnificent structures and mausoleums, adorn'd with statues and inscriptions, (planted about with cypress and other evergreens, and kept in repair) were not only graceful, but a noble and useful entertainment to the travellers, putting them in mind of the virtues and glorious actions of the persons buried ; of which I think, my Lord Verulam has somewhere spoken : However, there was certainly no permission for any to be buried within the walls of Rome, almost from the very foundation of it ; for so was the sanction, XII. tab. IN URBE NE SEPELITO NEVE URITO, Neither to bury or burn the dead in the city : And when long after they began to violate that law, Antonius Pius, and the Empp. succeeding, did again prohibit it : All we meet of ancient to the contrary, is of Cestius the Epulo's tomb, which is a thick clumsy pyramid, yet standing, *nec in urbe, nec in orbe;* as it were, but half in, and half without the wall. If then it were counted a thing so prophane to bury

in the cities, much less would they have permitted it
in their temples : Nor was it in use among Christians,
who in the primitive ages had no particular *coemeteria;*
but when (not long after) it was indulg'd, it was to
martyrs only *ad limina,* and in the porches, even to
the *deposita* of the [1] Apostles themselves.   Princes
indeed, and other illustrious persons, founders of
churches, &c. had sometimes their dormitories near
the basilica and cathedrals, a little before St. Augustine's
time ; as appears by his book *de cura pro mortuis,* and
the concession not easily obtain'd.   Constantine (son
to the Great Constantine himself) did not without
leave, inhume his royal father in the church-porch
of that august fabrick, tho' built by that famous
emperor;[2] and yet after this, other great persons
plac'd their sepulchres no nearer, than towards the
church-walls ; whilst in the body of the church, they
presum'd no farther for a long time after ; as may be
proved from the *Capitula* of Charle-Magni ; nor
hardly in the city, till the time of Gregory the Great;
and when conniv'd at, it was complained of : And
we find it forbidden (as to churches) by the emperors,
Gratian, Valentinian and Theodosius ; and so in the
code, where the sanction runs thus, *nemo apostolorum
vel martyrum sedem humanis corporibus existimet esse* [3]
concessam, &c.   And now after all this, would it not
raise our indignation, to suffer so many extortioners,
luxurious, profane, and very mean persons, without
merit, not only affecting, but permitted to lay their
carcasses, not in the nave and body of the church only,
but in the very chancel, next the Communion-table ;
ripping up the pavements, and removing the seats,

---

[1] So that passage of the famous civilian Baldwin, *ad leg.* XII. tab.
[2] Chrys. *Hom.* XXVI. *Epist. ad Corinth.*
[3] Gretzer, l. 2. *de fun. Christ.* c. 8. Onuphr. *de ritu sepul.*

&c. for some little gratification of those who should have more respect to decency at least, if for no other.

The fields, the mountains, the high-way-sides, and gardens, were thought enough honourable for those funeral purposes : Abraham and the Patriarchs (as we have shew'd) had their caves and crypta in the fields, set about with trees : The kings of Judah, their sepulchres in their palaces, not the Sanctuary and Temple: And our most Blessed Saviour's was in a garden ; which indeed seems to me to be the most proper and eligible, as we have already shew'd ; nor even to this day, do the Greeks and Eastern Christians bury in churches, as is well known.   A remarkable instance of this, we have of a worthy person of our own country : Mr. Burton, great grand-father of the learned author who writ the *Commentary* on Antoninus's *Itinerary;* which for its laudable singularity, I present my reader the description of: *In agro Salopiensi Lognorae ad Sabrinam, fl. ad piscinas in horto juxta aedes patruelis mei Francisci Burtoni pro-avi mei epitaphium;* with the following elegant title, 1558,

Quod scelus ? Aut Christi nomen temerare quod ausus,
   Hinc vetitum sacro condere membra solo ?
Di melius ; sincera fides, nec tramite veri
   Devia, causa ; illo tempore grande nefas.
Urbibus insultat nostris, dum turbida Roma ;
   Rasaque gens sacris dat sua jura locis :
Nec sacri ritus, nec honores funeris ; intra
   Moenia Christicolis, heu male sancta ! piis :
At referens Dominum inculpatae munera vitae,
   Ad Domini exemplar funera nactus erat.
Ille ut odorifero tumulatus marmore in horto :
   Ossa etiam redolens hortus & hujus habet,
Hic ubi & expectat, felix ! resonantia verba :
   Ergo age ! mercedem jam, bone serve, cape.

Thus with the incomparable Sannazarius ; *non mihi fornicibus Pariis.* Sculptures and titles preferrable to the proudest mausoleums I should chuse.

The late elegant and accomplished Sir W. Temple, tho' he laid not his whole body in his garden, deposited the better part of it (his heart) there ; and if my executors will gratify me in what I have desir'd, I wish my corps may be interr'd as I have bespoke them : Not at all out of singularity, or for want of a dormitory, (of which there is an ample one annext to the parish-church) but for other reasons, not here necessary to trouble the reader with ; what I have said in general, being sufficient : However, let them order it as they think fit, so it be not in the church or chancel.

Plato (as we noted) permitted trees to be planted over sepulchres, to obumbrate the departed : But with better reason; with flowers and redolent plants, emblems of the life of man, compar'd in Holy Scripture, to those fading beauties, whose roots being buried in dishonour, rise again in glory ; and of such hortulan instances, Greuter gives us this inscription,

> Hi horti ita uti opt. maximiq. sunt,
>   Cineribus serviant meis.
> Nam curatores substituam,
>   Qui vescantur
> Ex horum hortorum reditu
>   Natali meo.
> Et præbeant rosam in perpetuum.

This sweet flower, born on a branch full set with thorns, and accompany'd with the lilly, natural hieroglyphicks of our fugitive umbratile, anxious and transitory life, making so fair a shew for a time, is not without its thorn and crosses : These they therefore

planted on their turfy hillocks; like what is yet exstant in propylio D. Ambrosei a porto Vercelli.

PETRONIO JUCN VI. VIR.
SENI
PETRONIA MIRA L. F.
PATRONO QUÆ H. S.
Cccc LES POSSORIB
VICI BERDOMAS IN HERM.
TUENDO, ET ROSA QUOTANNIS
ORNANDUM.[1]

Of these and the like antiquity, we could multiply
instances, the custom not yet altogether extinct in my
own native county of Surrey, and near my dwelling ;
where the maidens yearly plant and deck the graves
of their defunct sweet-hearts with rose-bushes ; of
which I have given account in the learned Mr. Gib-
son's edition of *Camden;* and for the rest, see Mr. Sum-
ner, of *Garden-Burial,* and the learned Dr. Cave's
Primitive Christianity.

And now let not what I have said concerning the
pious Dr. Hammond's paraphrase in the text, of
hortulan burial, be thought foreign to my subject ;
since it takes in the custom of it in groves, and shady
and solemn places, as I have already shew'd ; and thus
the yew-trees at present growing, and planted in our
countrey church-yards, cypress, and other perennial
greens, emblems of immortality, and a reflourishing
state to come, were not less proper to shade our natural
beds, (would our climate suffer it) growing so like a
shrowd, as does that sepulchral tree.

To return then to that of groves, and for diversion

---

[1] There is a white amaracus, ᾧ χρῶνται περὶ τάφους used in funerals, v. Theoph.
de Plant. Athenæus, 1. 15. c. 7.

let us add a short recital of the most famous groves
which we find celebrated in histories; since those,
besides many already mention'd, were such as being
consecrated both to gods and men, bore their names.
Amongst these are reckoned the sacred to Minerva,
Isis, Latona, Cybele, Osiris, Æsculapius, Diana, and
especially the Aricinian, in which there was a goodly
temple erected, placed in the midst of an island, with
a vast lake about it, a mount, and a grotto adorn'd
with statues, and irrigated with plentiful streams :
And this was that renowned recess of Numa, where
he so frequently conversed with his Ægeria, as did
Minos in the Cave of Jupiter ; and by whose pretended
inspirations they gain'd the deceived people, and made
them receive what laws they pleas'd to impose upon
them. To these we may join the groves of Vulcan,
Venus, and the little youth Cupid;[1] Mars, Bellona,
Bacchus, Silvanus, the Muses, and that near Helicon
from the same Numa, their great patron ; and hence
had they their name Camœnæ. In this was the noble
statue of Eupheme nurse to those poetical ladies ; but
so the Feranian and even Mons Parnassus, were thick
shaded with trees. Nor may we omit the more
impure Lupercal groves, sacred, or prophan'd rather,
yet most famous for their affording shelter and foster
to Romulus, and his brother Remus.

That of Vulcan was usually guarded by dogs, like
the town of St. Malo's in Bretaigne : The *Pinea Silva*
appertain'd to the Mother of the Gods, as we find in
Virgil. Venus had several groves in Ægypt, and in
the Indian Island, where once stood those famous
statues cut by Praxiteles ; another in Pontus, where
(if you'll believe it) hung up the golden-fleece meed

[1] *Mars Silvanus* in ancient inscriptions, vide Catonem *de R. R.* c. XXXIII.

ee

of the bold adventurer. Nor was the watry-king Neptune without his groves, the Helicean in Greece was his: So Ceres, and Proserpine, Pluto, Vesta, Castor, and Pollux, had such shady places consecrated to them; add to these the Lebadian, Arsinoan, Paphian, Senonian, and such as were in general dedicated to all the Gods, for

[1] .... Gods have dwelt in groves.

And these were as it were Pantheons. To the memory of famous men and heroes were consecrated the Achillean, Aglauran, and those to Bellerophon, Hector, Alexander, and to others who disdained not to derive their names from trees and forests; as Silvius the *posthumus* of Æneas; divers of the Albanian Princes, and great persons; Stolon, Laura, Daphnis, &c. And a certain custom there was for the parents to plant a tree at the birth of an heir or son, presaging by the growth and thriving of the tree the prosperity of the child: Thus we read in the life of Virgil, and how far his natalitial poplar had out-stripp'd the rest of its contemporaries. And the reason doubtless of all this was, the general repute of the sanctity of those places; for no sooner does the poet speak of a grove, but immediately some consecration follows, as believing that out of those shady profundities, some Deity must needs emerge.

Quo possis viso dicere numen inest.

So as Tacitus (speaking of the Germans) says, *Lucos & nemora consecrant, deorumque nominibus appellant secretum illud, quod solâ reverentiâ vident;* To the same,

[1] .... Habitarunt di quoque silvas.

Pliny, l. 12. c. 1. *Arbores fuere numinum templa,* &c.
in which (says he) they did not so much revere the
golden and ivory statues, as the goodly trees and awful silence : And the consecration of these nemorous
places we find in Quintus Curtius, and in what Paulus
Diaconus relates of the Longobards, where the rites
are express, allur'd as 'tis likely by the gloominess of
the shade, procerity and altitude of the stem, floridness
of the leaves, and other accidents, not capable of
philosophizing on the physical causes, which they
deem'd supernatural, and plainly divine ; so as to use
the words of Prudentius,

[1] Here all religion paid ; whose dark recess
A sacred awe does on their mind impress,
To their wild gods.........

And this deification of their trees, and amongst other
things, for their age and perennial viridity, says Diodorus, might spring from the manifold use which they
afforded, and haply had been taught them by the gods,
or rather by some god-like persons, whom for their
worth, and the publick benefit they esteemed so; and
that divers of them were voic'd to have been meta-
morphoz'd from men into trees, and again out of trees
into men, as the Arcadians gloried in their birth, when

[2] Out of the teeming bark of oaks men burst,

which perhaps they fancied, by seeing men creep
sometimes out of their cavities, in which they often
lodg'd and secur'd themselves ;

[1] Quos penes omne sacrum est, quicquid formido tremendum
Suaserit horrificos, quos prodigialia cogunt
Monstra deos...... L. 2. *Cont. Sym.*
[2] Gensque virûm truncis & rupto robore nati.

¹ For in the earth's non-age under heaven's new frame,
They stricter liv'd who from oaks rupture came.

Stapylton.

Or as the sweet Papinius again,

> ² Fame goes that ye brake forth from the hard rind,
> When the new earth with the first feet was sign'd :
> Fields yet nor houses doleful pangs reliev'd,
> But shady ash the numerous births receiv'd,
> And the green babe dropt from the pregnant elm,
> Whom strange amazement first did over-whelm
> At break of day, and when the gloomy night
> Ravish'd the sun from their pursuing sight,
> Gave it for lost.........

almost like that which Rinaldo saw in the Inchanted Forest.

> ³ An aged oak beside him cleft and rent,
> And from his fertile hollow womb forth went
> (Clad in rare weeds, and strange habiliment)
> A full-grown nymph..........

And that every great tree included a certain tutelar genius or nymph living and dying with it, the poets are full ; a special instance we have in that prodigious oak which fell by the fatal stroke of Erisichthon;

---

¹ Quippe aliter tunc orbe novo caeloque recenti
Vivebant homines, qui rupto robore nati, &c.

Juvenal, vi. 11.

² ....Nemorum vos stirpe rigenti
Fama satos, cum prima pedum vestigia tellus
Admirata tulit, nondum arva, domusque ferebant,
Cruda puerperia, ac populos umbrosa creavit
Fraxinus, & fœtâ viridis puer excidit orno :
Hi lucis stupuisse vices, noctisque feruntur
Nubila, & occiduum longe Titana secuti
Desperàsse diem.........

³ Quercia gli appar, che perse stessa incisa
Apre feconda il cavo ventre, è figlia :
En'esce fuor vestita in strania guisa :
Ninfa d' eta cresciuta.......

Canto, 18.

but the Hamadryads it seems were immortal, and had power to remove and change their wooden habitations.

In the mean while, as to those nymphs (grieving to be dispossess'd of their ancient habitations) the fall of a very aged oak, giving a crack like thunder, has been often heard at many miles distance : Nor do I at any time, hear the groans without some emotion and pity ; constrain'd (as I too often am) to fell them with much reluctancy. Now that many such disasters have hap'ned to the owners of the places where goodly trees have been fell'd ; I cannot forget one, who giving the first stroke of the ax with his own hand, (and doubtless pursuing it with more) kill'd his own father by the fall of the tree, not without giving the uncautious Knight (for so he was) sufficient warning to avoid it. And here I must not pass by the groaning-board which they kept for a while in Southwark, drawing abundance of people to see the wonder ; such another plant being formerly, it seems, expos'd as a miracle at Caumont near Tolose in France, and as it sometimes happens in woods aud forests, thro' the inclusion of the air within the cavities of the timber ; and perhaps gave heretofore occasion of the fabulous Dodonian Oracle : But however it were, methinks I still hear, and am sure feel the dismal groans (hapning on the 26. Novemb. 1703.) of our forests, so many thousand of goodly oaks subverted by that late dreadful hurricane ; prostrating the trees, and crushing all that grew under them, lying in ghastly postures, like whole regiments fallen in battle, by the sword of the conqueror : Such was the prospect of many miles in several places, resembling that of Mount Taurus, so naturally describ'd by the poet, speaking of the fall of the Minotaurus slain by Theseus.

.........Illa procul radicitus exturbata,
Prona cadit, lateque et cominus obvia frangens.

The losses and dreadful stories of this ruin were
indeed great, but how much greater the universal devastation through the kingdom! The publick account
tells us, besides innumerable men, reckoning no less
than 3000 brave oaks, in one part only of the Forest
of Dean blown down ; and in New-Forest in Hamp-
shire about 4000; and in about 450 parks and groves,
from 200 large trees to a 1000 of excellent timber,
without counting fruit and orchard trees *sans* number,
and proportionably the same thorough all the considerable woods of the nation; with those stately groves,
avenues and vista's which the author names, especially
one tree of near eighty foot high, of clear timber 600
all subverted within the compass of five acres.

Sir Edward Harly had one thousand three hundred
blown down ; my self above 2000 ; several of which
torn-up by their fall, rais'd mounds of earth near 20 foot
high, with great stones intangled among the roots and
rubbish ; and this within almost sight of my dwelling,
(now no longer [1] Wotton) sufficient to mortifie and
change my too great affection and application to this
work ; which, as I contentedly submit to, so I thank
God for what are yet left standing : *nepotibus umbram.*

Lactantius reports of a people who worshipped the
wind, as some at this day among the Indians do the
Devil, that he may do them no harm.

What this prince of the air did to Job, and his religious family, for the trial of his patience, by God's
permission, the Scripture tells us : And for what cause
he still suffers that malicious Spirit to exert his fury in

---

[1] Wood-Town.

these lower regions, the same God only knows; though certainly for our chastisement ; and therefore reformation, submission and patience will become our best security.

Scaliger the father, affirms, he could never convince his learned antagonist Erasmus, but that trees felt the first stroke of the ax, and discovers a certain resentment: And indeed it seems to hold the edge of the fatal tool, till a wider gap be made : And so exceedingly apprehensive they are of their destruction, that as Zoröaster says, If a man come with a sharp bill, intending to fell a barren tree, and a friend importunately deprecate the angry person, and prevail with him to spare it, the tree will infallibly bear plentifully the next year: Such is the superstitious sanctity and folly of some credulous people.

But we were speaking of metamorphoses of one species into another ; as it is said of a platan into an olive-tree, when Xerxes came to Laodicea : And Lycosthenes talks of a *Sambucus* that bare grapes, which I believe he mistook for elder-berries.

Pliny mentions a timber-tree, that being felled, they found it full of stones, the solid wood grown over it: As it happened in Germany : Others (as above noted) that had armour, shields, and weapons invested with the timber of an old oak, which might have, when younger, been hung about it for trophies : But such another was found in Germany, that had the statue of the B. Virgin in the very centre of an aged oak of eight foot diameter, as John Burgosius affirms, and that the place where the tree stood was turned into a chappel near Dinand ad Mosam, famous for miracles: See his book *de parturitione B. M. Virg.*

15. We might here indeed produce the wonderful

strange apparitions of spirits interceding for the standing and life of trees, when the ax has been ready for
execution, as you may see in that hymn of Callimachus,
Pausanias[1], and the famous story of Parœbius related by
Apollonius in 2. *Argonaut.* with the fearful catastrophe
of such as causelesly and wantonly violated those goodly
plantations (from which fables arose that of the Do-
donean and vocal forests, frequent in heathen writers)
but by none so elegantly as the witty Ovid, describing
the fact of the wicked Erisichthon :

[2].........Who Gods despis'd,
Nor ever on their altars sacrific'd,
Who Ceres groves with steel prophan'd : where stood
An old huge oak ; even of it self a wood.
Wreaths, ribbands, grateful tables deckt his boughs
And sacred stem ; the dues of powerful vows.
Full oft the Dryades, with chaplets crown'd,
Danc'd in the shade, full oft they tript a round
About his bole.   Five cubits three times told
His ample circuit hardly could infold.
Whose stature other trees as far exceeds,
As other trees surmount the humble weeds.
Yet this his fury rather did provoke :
Who bids his servants fell the sacred oak.
And snatches, while they paus'd, an ax from one,
Thus storming : Not the Goddess lov'd alone ;
But, though this were the Goddess, she should down,
And sweep the earth with her aspiring crown.
As he advanc'd his arms to strike, the oak
Both sigh'd and trembl'd at the threatning stroke ;
His leaves and acorns, pale together grew,
And colour-changing branches sweat cold dew :
Then wounded by his impious hand, the blood
Gush'd from the incision in a purple flood :

[1] In Phoe. & Arcad.
[2] ....Qui numina divom
Sperneret, & nullos aris adoleret honores, &c.

Much like a mighty ox, that falls before
The sacred altar, spouting streams of gore.
On all amazement seiz'd : When one of all
The crime deters, nor would his ax let fall.
Contracting his stern brows ; Receive, said he,
Thy pieties reward ; and from the tree
The stroke converting, lops his head ; then strake
The oak again ; from whence a voice thus spake :
A nymph am I, within this tree inshrin'd,
Belov'd of Ceres, O prophane of mind,
Vengeance is near thee : with my parting breath,
I prophesie, a comfort to my death.
He still his guilt pursues ; who over-throws
With cables, and innumerable blows,
The sturdy oak ; which nodding long, down rush'd,
And in his lofty fall his fellows crush'd.

Sandys.

But a sad revenge follows it, as the poet will tell
you ; and one might fill a just volume with the
histories of groves that were violated by wicked men,
who came to fatal periods ; especially those upon
which the misselto grew, than which nothing was
reputed more sacred,

[1] To misselto the Druids us'd to sing.

For among such oaks they usually dwelt,

.... Nemora alta remotis
Incolitis lucis..........
Lucan.

with whose leaves they adorn'd and celebrated their
religious rites. The Druids, says Pliny, lib. 16. c. 4.
(for so they call their divines) esteem nothing more

---

[1] Ad viscum Druidae, Druidae cantare solebant.

venerable than misselto, and the oak upon which it grows, &c. Indeed they did nothing of importance, without some leaves or branches of this tree, and its very excrescence as sent from heaven, and with a solemn sacrifice of two white bulls ; the misselto not to be gather'd, but cut by the priest with a golden ax, praying for a blessing on this divine gift, &c. But of this consult (besides the author) Mela, Lactantius, Eusebius *de praeparat. evangel.* and the *Aulularia* of Pseudo-Plautus, Camden and others ; whilst as to that excrescence, I am told of the disasters which happened to the two men who (not long since) fell'd a goodly tree, call'd the Vicar's Oak, standing at Nor-Wood (not far from Croydon) partly belonging to the arch-bishop, and was limit to four parishes, which met in a point ; on this oak grew an extraordinary branch of misselto, which in the time of the sacriligious usurpers they were wont to cut and sell to an apothe-cary of London; and though warn'd of the misfortunes observed to befall those who injured this plant, proceeding not only to cut it quite off, without leaving a sprig remaining, but to demolish and fell the oak it self also : The first soon after lost his eye, and the other brake his leg ; as if the Hamadryads had revenged the indignity.

It is reported that the Minturnensian grove was esteem'd so venerable, that a stranger might not be admitted into it ; and the great Xerxes himself, when he passed through Achaia, would not touch a grove which was dedicated to Jupiter, commanding his army to do it no violence ; and the honours he did to one single (but a goodly) *platanus*, we have already mentioned. The like to this we find when the Per-sians were put to flight by Pausanias ; though they

might have sav'd their lives by it, as appears in the
story. The same reverence made that Hercules
would not so much as tast the waters of the Ægerian
groves, after he slew Cacus, though extremely thirsty.

[1] .... The priestess said
(A purple fillet binding her gray head)
Stranger, pry not, but quit this shady seat,
Avant, and whiles thou safely may'st, retreat,
To men forbid, and by hard sanction bound :
Far better other springs were by thee found.

Nor indeed in such places was it lawful to hunt,
unless it were to kill for sacrifice, as we read in
Arrianus ; whence 'tis reported by Strabo, that in the
Ætolian groves sacred to Diana, the beasts were so
tame, that the very wolves and stags fed together like
lambs, and would follow a man licking his hands, and
fawning on him. Such a grove was the Crotonian,
in which Livy writes, there was a spacious field like
St. James's Park, stored with all sorts of game. There
were many forests consecrated to Jupiter, Juno, and
Apollo ; especially the famous Epidaphne, near the
Syrian Antioch, which was most incomparably pleasant, and adorn'd with fountains and rare statues.
[2] There was to be seen the laurel which had been his
chast mistress, and in the centre of it his temple, an
asylum : Here it was Cosroes and Julian did sacrifice
upon several occasions, as Eusebius relates, but could
not with all their impious arts obtain an answer ;

[1] Puniceo canas stamine vincta comas :
Parce oculis, hospes, lucoque abscede verendo,
    Cede agedum, & tuta limina linque fuga
Interdicta viris ; metuenda lege piatur,
    Quae se summota vindicat ara casa.....
Di tibi dent alios fontes............   *Propert.* iv. 9. 52.

[2] See this delicious place elegantly described by S. Chrysostom, *Lib. de
S. Babyl.* Tom. VI. p. 671. Sozom. Lib. VI. cap. 19. Niceph. Lib. X. cap. 28.

because the holy Babylas had been interr'd near that oracle ; for which it was reputed so venerable, that there remained an express title in the code, *de cupressis ex luco Daphnes non excidendis, vel venundandis*, that none should either fell, or sell any of the trees about it ; which may serve for another instance of their burying in such places.   The truth is, so exceedingly superstitious they were and tender, that there was almost no medling with these devoted trees, and even before they did but *conlucare* and prune one of them, they were first to sacrifice lest they might offend in something ignorantly : But to cut down was capital, and never to be done away with any offering what-soever ; and therefore *conlucare* in authors, is not (as some pretend) *succidere*, but to prune the branches only ; and yet even this gentle tonsure of superfluities was reputed a kind of contamination ; and hence *lucus coinquinari dicitur*, unless in the case of lightning, when *caelo tacti*, a whole tree might quite be felled, as marked, by heaven for the fire : But of this suffi-cient.   We could indeed fill many sheets with the catastrophe of such as maliciously destroy'd groves, to feed either their revenge or avarice : See Plutarch in *Pericles*, and the saying of Pompeius: Cicero sharply reproves C. Gabinius for his prodigious spoil in Greece; and it was of late days held a piece of inhumanity in Charles, the French king, when he entred the Frisons after he had slain their leader, to cut down their woods, a punishment never inflicted by sober princes, but to prevent idolatry in the old law; and to shew the heinousness of disloyalty and treason by latter sanctions; in which case, and for terror, even a traytor's woods have become anathema, as were easie to instance out of histories.

10. But what shall we say then of our late prodigious spoilers, whose furious devastation of so many goodly woods and forests, have bequeath'd an infamy on their names and memories not quickly to be forgotten! I mean our unhappy usurpers, and injurious sequestrators; not here to mention the deplorable necessities of a gallant and loyal gentry, who for their compositions were (many of them) compelled to add yet to this wast, by an inhumane and unparallel'd tyranny over them, to preserve the poor remainder of their fortunes, and to find them bread.

Nor was it here they desisted, when, after the fate of that once beautiful grove under Greenwich-Castle, (of late supply'd by his present Majesty) the Royal walk of elms in St. James's Park,

That living gallery of aged trees,

was once propos'd to the late Council of State (as they called it) to be cut down and sold, that with the rest of his Majesty's houses already demolished, and marked out for destruction, his trees might likewise undergo the same destiny, and no footsteps of monarchy remain unviolated.

17. It is from hence you may culculate what were the designs of those excellent reformers, and the care these great states-men took for the preservation of their country, when being parties in the booty themselves, they gave way to so dishonourable and impolitic a wast of that material, which being left entire, or husbanded with discretion, had proved the best support and defence of it. But this (say they) was the effect of war, and in the height of our contentions. No, it was a late and cold deliberation, and long after all had been subdued to them; nor could the most

implacable of enemies have express'd a resolution
more barbarous.

For, as our own incomparable poet describes it,

> ......'Twas not enough alone to take the spoils
> Of God's, and the king's houses ; these unjust
> And impious men destroy the stately piles :
> Of very ruin there's a wicked lust.

> In every place the groaning carts are fill'd
> With beams and stones, so busie and so loud
> Are the proud victors, as they meant to build,
> But they to ruin and destruction crowd.

> Timber, which had been buried many years
> Under such royal towers, they invade :
> 'Tis sure that hand the living never spares,
> Which is so wicked to disturb the dead.

> Then all the woods the barbarous victors seize,
> (The noble nursery of the fleet and town,
> The hopes of war, and ornaments of peace)
> Which once religion did as sacred own.

> Now publick use, and great convenience claims,
> The woods from private hands inviolate ;
> Which greedy men to less devouring flames,
> Do for sweet lucre freely dedicate.

> No age they spare, the tender elm and beech,
> Infants of thirty years they overthrow ;
> Nor could old age it self their pity reach,
> No reverence to hoary barks they know.

> Th' unhappy birds, an ever-singing quire,
> Are driven from their ancient shady seats,
> And a new grief does Philomel inspire
> With mournful notes, which she all night repeats.

Let them the woods and forest burn and waste,
There will be trees to hang the slaves at last,
And God, who such infernal men disclaims,
Will root 'em out, and throw' em into flames.

In which he has shew'd himself as well a prophet
as a poet.

We have spoken of the great Xerxes, that passing
conqueror through Achaia, he would not suffer his
army to violate so much as a tree of his adversaries ;
and have sufficiently observed from the ancients, that
the [1] Gods did never permit them to escape unpunish'd
who were injurious to groves. What became of
Agamemnon's host after his spoil of the woods at
Aulis ? Histories tell us Cleomenes died mad : The
Temesæan Genius became proverbial; and the destructive fact that the inraged Cæsar perpetrated on the
Massilian trees, went not long unrevenged ; thus
related by the poet, and an illustrious record of all
we have hitherto produc'd to assert their veneration:

Lucus erat longo nunquam violatus ab ævo, &c.

Lucan. l. 3.

A wood untouch'd of old was growing there
Of thick-set trees whose boughs spreading and fair,
Meeting, obscured the inclosed air,
And made dark shades exiling Phœbus rayes :
There no rude Fawn, nor wanton Silvan plays ;
No nymph disports, but cruel deities
Claim barbarous rites, and bloody sacrifice :
Each tree defil'd with human blood ; if we
Believe traditions of antiquity :
No bird dares light upon those hallowed boughs,

---

[1] Though cut down for building of ships.
Lucum Aesculapio dicatum succiderat Turullius: manifestis numinis illius viri-
bus, eum in lucum quem violaverat, ille attractus est, effecitque deus ut ibi potis-
simum occideretur. Vide Valer. Max. lib. I. cap. I. n. 19.

No beasts make there their dens ; no wind there blows ;
No lightning falls : A sad religious awe,
The quiet trees unstirr'd by wind do draw.
Black water currents from dark fountains flow ;
The Gods unpolish'd images do know
No art, but plain, and formless trunks they are.
Their moss and moldiness procures a fear :
The common figures of known deities
Are not so sear'd : Not knowing what God 'tis,
Makes him more awful ; by relation
The shaken earth's dark caverns oft did groan :
Fallen yew-trees often of themselves would rise :
With seeming fire oft flam'd th' unburned trees :
And winding dragons the cold oaks embrace,
None give near worship to that baleful place ;
The people leave it to the Gods alone.
When black night reigns, or Phœbus guilds the noon,
The priest himself trembles, afraid to spy
In th' awful woods its guardian deity.

But now Erisichthon-like, and like him in punishment ; for his was hunger, Cæsar's thirst, and thirst
of human blood, reveng'd soon after in his own.

The woods he bids them fell, not standing far
From all their work : Untouch'd in former war
Among the other bared hills it stands
Of a thick growth ; the soldiers valiant hands
Trembled to strike, mov'd with the majesty,
And think the ax from off the sacred tree
Rebounding back, would their own bodies wound :
Th' amazement of his men when Cæsar found,
In his bold hand himself an hatchet took,
And first of all assaults a lofty oak ;
And having wounded the religious tree,
Let no man fear to fell this wood (quoth he)
The guilt of this offence let Cæsar bear, &c.

May.

and so he did soon after, carrying ('tis thought) the
maledictions of the incensed Gauls to his funeral pile,

[1] .... For who
The Gods thus injur'd unreveng'd does go ?

18. But lest this be charged with superstition,
because the instances are heathen ; it was a more noble
and remarkable, as well as recent example, when at
the siege of Breda, the late famous general Spinola
commanded his army not to violate a tree of a certain wood belonging to the Prince of Orange there,
tho a reputed traytor, and in open defiance with his
master.   In sum, we read that when Mithridates but
deliberated about the cutting down of some stately
trees which grew near Patara, a City of Lycia, tho
necessitated to it for the building of warlike engines
with them, being terrified in a vision, he desisted
from his purpose.   It were to be wished these, or the
like examples, might have wrought some effects
upon the sacrilegious purchasers, and disloyal invaders
in this iron-age amongst us, who have lately made
so prodigious a spoil of those goodly forests, woods,
and trees (to gratifie an impious and unworthy
avarice) which being once the treasure and ornament
of this nation, were doubtless reserved by our more
prudent ancestors for the repairs of our floating
castles, the safeguard and boast of this renowned
island, when necessity, or some imminent peril should
threaten it, or call for their assistance ; and not to be
devoured by these improvident wretches, who, to
their eternal reproach, did (with the royal patrimony)

[1] .... Quis enim læsos impune putaret
Esse deos ?

gg

swallow likewise God's [1] own inheritance ; but whose
sons and grand-children we have lived to see as
hastily disgorge them again ; and with it all the rest
of their holy purchases, which otherwise they might
securely have enjoyed. But this, *in terrorem* only,
and for caution to posterity, whiles we leave [2] the
guilty, and those who have done the mischiefs, to
their proper scorpions, and to their Erisichthonian-
fate, or that of the inexorable Parœbius, the vengeance
of the Dryads, and to their tutelar better genius, if
any yet remain, who love the solid honour and
ornament of their country : For what could I say
less, tho constrained by necessity my self, to cut down
so many goodly trees, and venerable woods, (devoted
to the ax by the owner, who had right to dispose of
them before me) ʽΥλογενὴς, and [3] wood-born as I am,
in behalf of those sacred shades, which both grace
our habitations, and protect our nation ? So in all
ages, from trees have been denominated whole
countries, regions, cities and towns ; as Cyparissa in
Greece, Cerasus in Pontus, Laurentum in Italy,
Myrrhinus in Attica.   Ports, mountains, and eminent
places ; as the Viminalis, Æsculetum, &c. The
reason is obvious, from the spontaneous growth and
abounding of such trees in the respective soils : And
hence of old, *Avellana nux*, is called also *Praenestina,*
Pontica ; dum unaquaeque natio indit huic nuci nomen ex
loco in quo nascitur copiosior : So the chesnut, called
Heracleotica, of which see Macrob. *Saturnal.* l, 3.
And Sylvius became great and famous names among
the Latines and Romans : Sylvius Posthumus, the

---

[1] Quæ tibi factorum pœnas instare tuorum vaticinor.....
[2] Vide Met. l. 8. Apollon. l. 2. *Argonaut.* Prosternit quercum funestam quam
sibi nympha pignoribusque suis fecit........
[3] At Wotton in Surrey.

son of that renowned hero Æneas Sylvius ; and in
time an hereditary name among the subsequent kings:
Latinus Sylvius, Alba Sylvius, who built that glorious
city, which contended with Rome her self : And to
return to our own country, Seven-oaks in Kent was
so called (as reported) from some goodly oaks growing
about it, and giving name also to that Lord Mayor
(a foundling of that place) and was himself the
founder of the first protestant hospital in England,
defeated the insurrection of J. Cade, and his complices,
for which he was knighted, as he deserved.

Old Sarum, or Sorbiodunum, had its name *a sorbis.*

Hence also from the plenty of beech-trees does
Mr. Camden denominate the whole county of Buck-
ingham, Bukenham in Norfolk, Buchonia in Germany,
&c. though indeed the learned author of the additions
to the late edition, think them rather so called from
the Saxon *buc (cervus)* a buck, or hart, and this from
that in Norfolk, where Sir Henry Spelman reports
there are no such trees growing ; whilst we yet know
not whether there may not formerly have been store:
In all events, be it one or the other, it is certain,
abundance of places, countries and families have taken
their denomination from trees.

One thing more I think not impertinent to hint,
before I take my leave of this book, concerning the use
of standing groves ; that in some places of the world,
they have no other water to drink than what their trees
afford them; not only of their proper juice (as we have
noted) but from their attraction of the evening moisture,
which impends in the shape of a cloud over them :
Such a tufft of trees is in the island of Ferro, of which
consult the learned Isaac Vossius upon Pomponius
Mela, and Magnenus de Manna : The same likewise

hapning in the Indies ; so that if their woods were once destroyed, they might perish for want of rains ; upon which account Barbadoes grows every year more torrid, and has not near the rain it formerly enjoyed when it was better furnished with trees ; and so in Jamaica at Gunaboa, the rains are observed to diminish, as their plantations extend : The like I could tell you of some parts of England not far from hence.

And now lastly, to encourage those to plant that have opportunity, and those who innocently, and with reluctancy are forced to cut down, and endeavour to supply the waste with their utmost industry : 'Tis observed that such planters are often bless'd with health and old age, according to that of the prophet LX *Isa.* 22 : " The days of a tree are the days of my people ": Instances of whose extraordinary longœvity, we have given abundance in this discourse, and seems to be so universally remarked, that as Paulus Venetus (that great traveller) reports, the Tartarian astrologers affirm, nothing contributes more to mens long lives, than the planting of many trees : *Haec scripsi octogenarius,* and shall, if God protract my years, and continue health, be continually planting, till it shall please him to transplant me into those glorious regions above, the cœlestial Paradise, planted with perennial groves and trees, bearing immortal fruit ; for such is the tree of life, which they who do his commandments have right to, XXII *Apoc.* 20 : Ναὶ, ἔρχομαι ταχύ, ἀμήν ναὶ, ἔρχου Κύριε Ἰησοῦ ἀμήν.

19. Thus my reader sees, and I acknowledge, how easie it is to be lost in the woods, and that I have hardly power to take off my pen whilst I am on this delightful subject : For what more august, more

charming and useful, than the culture and preservation
of such goodly plantations,

> [1] That shade to our grand-children give ?

and afford so sweet, and so agreeable refreshment to
our industrious wood-man,

> [2] When he his wearied limbs has laid
> Under a florid platan's shade.

or some other goodly spreading trees, such as we told
you stopt the legions of a proud conqueror, and that
the wise Socrates sware by ; that Passienus Crispus did
sacrifice to, and the honours of his gods ?

20. But whilst we condemn this excess in them,
Christians and true philosophers may be instructed to
make use of these enjoyments to better purposes, by
contemplating the miracles of their production and
structure : And what mortal is there so perfect an
atomist, who will undertake to detect the thousandth
part, or point of so exile a grain, as that insensible
rudiment, or rather halituous spirit, which brings forth
the lofty fir-tree, and the spreading oak ? that trees of
so enormous an height and magnitude, as we find
some elms, planes, and cypresses ; some hard as iron,
and solid as marble (for such the Indies furnish many)
should be swadl'd and involv'd within so small a
dimension (if a point may be said to have any) without
the least luxation, confusion or disorder of parts, and
in so weak and feeble a substance; being at first but a
kind of tender mucilage, or rather rottenness, which

---

[1] .... Seris factura nepotibus umbram.

[2] Cum post labores sub platani cubat
   Virentis umbra.........

                                        Claud.

so easily dissolves and corrupts substances so much harder, when they are buried in the moist womb of the earth, whilst this tender and flexible as it is, shall be able in time to displace and rend in sunder whole rocks of stones, and sometimes to cleave them beyond the force of iron wedges, so as even to remove mountains? For thus no weights are observed able to suppress the victorious palm : And thus our tree (like man whose inverted symbol he is) being sown in corruption, rises in glory, by little and little ascending into an hard erect stem of comely dimensions, into a solid tower, as it were; and that which but lately a single ant would easily have born to his little cavern, now capable of resisting the fury, and braving the rage of the most impetuous storms, *magni mehercie artificis, clausisse totum in tam exiguo* (to use Seneca's expression) *& horror est consideranti.*[1]

For is it not plainly astonishing how these minute atoms, rather than visible eggs, should contain the fœtus exquisitely formed, even while yet wrap'd in their secondines, like infants in the animal womb, till growing too big for the dark confinements, they break forth, and after a while more distinctly display every limb and member compleatly perfect, with all their apparel, tire and trim of beautiful and flourishing vegetables, endow'd with all the qualities of the species.

21. Contemplate we again, what it is which begins the motion, and kindles the flame of these *automata*, causing them first to radiate in the earth, and then to display their top in the air, so different poles, (as I may call them) in such different mediums; what it is imparts this elastic, peristaltic and other motions, so

[1] *Epist.* 53.

very like to the sensible, and perfectest animal; how
they elect, and then intro-sume their proper food, and
give suck, as it were, to the yet tender infant, till it
have strength and force to prey on, and digest the
more solid juices of the earth; for then, and not till
then, do the roots begin to harden: Consider how they
assimilate, separate and distribute these several supplies; how they concoct, transmute, augment, produce
and nourish without separation of excrements (at least
to us visible) and generate their like, whilst furnished
with tubes, ovaries, umbilical and other vessels, the
principle of any species, are safely reserved and nourished till delivered without violation of virginity: By
what exquisite percolations and fermentations they
proceed: for the heart, fibers, veins, nerves, valves and
anastomotas, rind, branches, leaves, blossoms, fruit; for
the strength, colour, taste, odour and other stupendous
qualities, and distinct faculties, some of them so re-
pugnant and contrary to others; yet in so uniform and
successive a series, and all this performed in the dark,
and those secret recesses of nature: With what [1] ana-
logy the solider and inflexible texture of parts of trees
agree with the bones, ribs, vertibræ, &c. nay, with the
very brains and marrow, and the more pliable, fitted to
such various motions, have induced some to allow them
place among the class of animals, is astonishing: To
these, and for their preservation, nature has invested
the whole tribe and nation (as we may say) of vegetables, with garments suitable to their naked and
exposed bodies, temper and climate: Thus some are
clad with a courser, and resist all extremes of weather;
others with more tender, and delicate skins and scarfs

[1] See Scaliger *Exerc.* 14. of respondent parts, within and without, from head to foot.

as it were, and thinner rayment. *Quid foliorum descri-
bam diversitates ?* what shall we say of the mysterious forms, variety and variegation of the leaves and
flowers, contrived with such art, yet without art ;
some round, others long, oval, multangular, indented,
crisped, rough, smooth and polished, soft and flexible
at every tremulous blast, as if it would drop in a
moment, and yet so obstinately adhering, as to be able
to contest against the fiercest winds, that prostrate
mighty structures, resisting hurricanes, the violence
whereof whole fleets and countries do often feel; yet
I say, continually making war, and sometimes joining
forces with steeming showers, against the poor leaf,
tied on by a slender stalk ! there it abides till God
bids it fall: For so the wise Disposer of things has
plac'd it, not only for ornament, but use and protection
both of body and fruit, from the excessive heat of
summer, and colds even of the sharpest winters, and
their immediate impressions; as we find it in all such
places and trees, as like the blessed and good man,
have always fruit upon them, ripe, or preparing to
mature; such as the pine, fir, arbutus, orange, and most
of those which the Indies and more southern tracts
plentifully abound in, where nature provides this
continual shelter, and cloaths them with perennial
garments.

But with what amazement do we consider what
may be demonstrated of the innumerable (and next to
infinite) number of seeds, which in a young elm (for
instance) it would amount, during the ordinary age
of that species, which suppose to be but one hundred
years standing, it has in it 15480000000 seeds, and
the tree grow and multiply, as many times, every
individual grain contain a second tree, including the

like number, and so on by geometrical progression in squares and cubes, &c. At what a loss must the most enlarged human capacity be at so stupendous a consideration !

One single seed of tobacco would produce 1296000 000000000, &c. and every one of these how many more, let those who have leisure compute.

22. Let us again examine with what care the seeds, those little souls of plants, *Quorum exilitas* (as one says) *vix locum inveniat* (in which the whole and compleat tree, though invisible to our dull sense, is yet perfectly and entirely wrapp'd up) are preserved from avolation, diminution aud detriment ; expos'd, as they seem to be, to all those accidents of weather, storms, and rapacious birds, in their spiny, arm'd and compacted receptacles ; where they sleep as in their causes, till their prisons let them gently fall into the embraces of the earth, now made pregnant with the season, and ready for another burthen : For at the time of year she fails not to bring them forth. And with what delight have I beheld this tender and innumerable off-spring, repullulating at the feet of an aged tree ! from whence the suckers are drawn, transplanted and educated by human industry, and forgetting the ferity of their nature, become civiliz'd to all his employments.

23. Can we look on the prodigious quantity of liquor, which one poor wounded birch will produce in a few hours, and not be astonished how some trees should in so short a space, weep more than they weigh ? And that so dry, so feeble and wretched a branch, as that which bears the grape, should yield a juice that cheers both God and man ? That the pine, fir, larch, and other resinous trees, planted in such

hh

rude and uncultivated places, amongst rocks and dry
pumices, should transude into turpentine, and pearl
out into gums, and precious balms ?

In a word, so astonishing and wonderful is the organ-
isms, parts and functions of plants and trees ; as some
have, as we said, attributed animal life to them, and
that they were living creatures; for so did Anaxagoras,
Empedocles, and even Plato himself.

I am sure plants and trees afford more matter for [1]
medicine, and the use of man, than either animals and
minerals, or any exotic we have besides ; are more
familiar at hand, and safe ; and within this late age
wonderfully improved, increased and searched into,
and seems by the Divine wisdom, to be an inexhaust-
ible subject for our disquisition and admiration.

24. There are ten thousand considerations more,
besides that of their medicinal and sanative properties,
and the mechanical uses mentioned in this treatise,
which a contemplative person may derive from the
groves and woods ; all of them the subject of wonder :
And though he had only the palm, (which [2] Strabo
affirms is fit for three hundred and sixty uses ;) or the
coco, which yields wine, bread, milk, oyl, sugar,
vinegar, tinctures, tanns, spices, thread, needle, linnen,
and cloth, cups, dishes, spoons, and other vessels and
utensils ; baskets, mats, umbrellas, paper, brooms,
ropes, sails, and almost all that belongs to the rigging
of ships. In short, this single tree furnishing a great
part of the world with all that even a voluptuous man
can need, or almost desire ; it were sufficient to employ his meditations and his hands, as long as he were
to live, though his years were as many as the most

[1] *Vide* Petri Mangot *Botan. Monspel.*
[2] *Vide* Mr. Dodart's *Hist. de l'Academ. Scient.*

aged oak :   So as Fr. Hernander, Gracilasco de la
Urga, and other [1] travellers, speaking of the coco,
aloes, wild-pine of Jamaica, &c. affirm there is nothing
necessary for life (*si effet rebus humanis modus*), which
these polychrests afford not.

What may we say then of innumerable other trees,
fitted for the uses nature has designed them, especially
for timber, and all other fabrile employments ?   But
I cease to expatiate farther on these wonders, that it
may not anticipate the pleasures which the serious
contemplator on those stupendous works of nature,
(or rather God of nature) will find himself even rapt'd
and transported, were it only applied to the production of a single wood.

Let the further curious, or those who may take
these wonders for a florid *epiphonema* only of this
work ;   add to the most ancient naturalists, what they
will find improved on this ample subject, in the late
excellently learned and judicious Malphigius, Grew,
Ray, Senertus, Faber, and others who have defin'd
these astonishing operations of nature, causes and
effects, with the greatest and exactest ἀκρίβεια imaginable.   But a wise and a thinking man can need
none of these topics ; in every hedge, and every field
they are before him; and yet we do not admire them
because they are common and obvious : Thus we
fall into the just reproach given by one of the philosophers (introduced by the Orator), [2] to those who
slighted what they saw every day, because they every
day saw them: *Quasi novitas nos quom magnitudo rerum,*
debeat ad exquirendas causas excitare: As if novelty only
should be of more force to engage our enquiry into

[1] *Vide* Ray *H. Pl. L.* XXI. C. 7.
[2] Cic. *de Nat. Deor.* l. 2.

the causes of things, than the worth and magnitude of the things themselves.

I conclude this book, and whole discourse with that incomparable poem of Rapinus, as epitomizing all we have said.

I cannot therefore but wonder, that excellent piece, (so elegant, pleasant and instructive) should be no more enquired after.

RENATI RAPINI. S. J. HORTORUM
Lib. II. NEMUS.

Me nemora, atque omnis nemorum pulcherrimus ordo
Et spatia, umbrandum late fundenda per hortum
Invitant, &c.

Thus made English by my late son Evelyn.

Long rows of trees and woods my pen invite,
With shady walks a garden's chief delight :
For nothing without them is pleasant made :
They beauty to the ruder country add.
Ye woods and spreading groves afford my muse
That bough, with which the sacret poets use
T' adorn their brows ; that by their pattern led,
I with due laurels may impale my head.

Methinks the oaks their willing tops incline,
Their trembling leaves applauding my design ;
With joyful murmurs, and unforc'd assent,
The woods of Gaule accord me their consent.
Cithaeron I, and Menalus despise,
Oft grac'd by the Arcadian deities ;
I, nor Molorchus, or Dodona's grove,
Or thee crown'd with black oaks, Calydne love ;
Cyllene thick with cypress too I fly ;
To France alone my genius I apply,
Where noble woods in ev'ry part abound,
And pleasant groves commend the fertile ground.

If on thy native soil thou dost prepare
T' erect a villa, you must place it there,
Where a free prospect do's it self extend
Into a garden whence the sun may lend
His influence from the east ; his radiant heat
Should on your house through various windows beat ;

But on that side which chiefly open lies
To the north-wind, whence storms and show'rs arise,
There plant a wood; for, without that defence,
Nothing resists the northern violence.
While with destructive blasts o're clifts and hills
Rough Boreas moves, and all with murmurs fills;
The oak with shaken boughs on mountains rends,
The valleys roar, and great Olympus bends.
Trees therefore to the winds you must expose,
Whose branches best their pow'rful rage oppose.

Thus woods defend that part of Normandy,
Which spreads it self upon the British Sea.
Where trees do all along the ocean side
Great villages and meadows too divide.

But now the means of raising woods I sing;
Tho from the parent oak young shoots may spring,
Or may transplanted flourish, yet I know
No better means than if from seed they grow.
'Tis true this way a longer time will need,
And oaks but slowly are produc'd by seed:
Yet they with far the happier shades are blest;
For those that rise from acorns, as they best
With deep-fixt roots beneath the earth descend,
So their large boughs into the air ascend.
Perhaps because, when we young sets translate,
They lose their virtue, and degenerate,
While acorns better thrive, since from their birth
They have been more acquainted with the earth.

Thus we to woods by acorns being give;
But yet before the ground your seed receive,
To dig it first employ your labourer;
Then level it; and, if young shoots appear
Above the ground, sprung from the cloven bud;
If th' earth be planted in the spring, 'tis good
Those weeds by frequent culture to remove,
Whose roots would to the blossom hurtful prove.
Nor think it labour lost to use the plow;
By dung and tillage all things fertile grow.

There are more ways than one to plant a grove,

For some do best a rude confusion love ;
Some into even squares dispose their trees,
Where ev'ry side do's equal bounds possess.
Thus boxen legions with false arms appear
At chess, and represent a face of war.
Which sport to *schaccia* the Italians owe ;
The painted frames alternate colours show.
So should the field in space and form agree ;
And should in equal bounds divided be.

Whether you plant young sets, or acorns sow,
Still order keep ; for so they best will grow.
Order to ev'ry tree like vigour gives,
And room for the aspiring branches leaves.

When with the leaf your hopes begin to bud,
Banish all wanton cattle from the wood.
The browzing goat the tender blossom kills ;
Let the swift horse then neigh upon the hills,
And the free herds still in large pastures tread ;
But not upon the new-sprung branches feed.
For whose defence inclosures should be made
Of twigs, or water into rills convey'd.
When ripening time has made your trees dilate,
And the strong roots do deeply penetrate,
All the superfluous branches must be fell'd,
Lest the oppressed trunk should chance to yield
Under the weight, and so its spirits lose
In such excrescencies ; but as for those
Which from the stock you cut, they better thrive,
As if their ruin caus'd them to revive.
And the slow plant, which scarce advanc'd its head,
Into the air its leavy boughs will spread.

When from the fastned root it springs amain,
And can the fury of the north sustain ;
On the smooth bark the shepherds should indite
Their rural strifes, and there their verses write.

But let no impious ax prophane the woods,
Or violate the sacred shades ; the Gods
Themselves inhabit there.   Some have beheld
Where drops of blood from wounded oaks distill'd :

Have seen the trembling boughs with horror shake !
So great a conscience did the ancients make
To cut down oaks, that it was held a crime
In that obscure and superstitious time,
For Dryopeius heaven did provoke,
By daring to destroy th' Æmonian oak ;
And with it it's included Dryad too :
Avenging Ceres here her faith did show
To the wrong'd nymph ; while Erisichthon bore
Torments, as great as was his crime before.
Therefore it well might be esteem'd no less
Than sacriledge, when ev'ry dark recess,
The awful silence, and each gloomy shade,
Was sacred by the zealous vulgar made.
When e're they cut down groves, or spoil'd the trees,
With gifts the ancients Pales did appease.

Due honours once Dodona's forest had,
When oracles were through the oaks convey'd.
When woods instructed prophets to foretell,
And the decrees of fate in trees did dwell.

If the aspiring plant large branches bear,
And beeches with extended arms appear ;
There near his flocks upon the cooler ground
The swain may lie, and with his pipe resound
His loves ; but let no vice these shades disgrace :
We ought to bear a rev'rence to the place.
The boughs, th' unbroken silence of a wood,
The leaves themselves demonstrate that some God
Inhabits there, whose flames might be so just,
To burn those groves that had been fir'd by lust.

But through the woods while thus the rusticks sport,
Whole flights of birds will thither too resort ;
Whose diff'rent notes and murmurs fill the air :
Thither sad Philomela will repair ;
Once to her sister she complain'd, but now
She warbles forth her grief on ev'ry bough :
Fills all with Tereus crimes, her own hard fate ;
And makes the melting rocks compassionate.
Dusturb not birds which in your trees abide,
By them the will of heav'n is signify'd :

How oft from hollow oaks the boading crow,
The winds and future tempests do's foreshow !
Of these the wary plowman should make use ;
Hence observations of his own deduce :
And so the changes of the weather tell.
But from your groves all hurtful birds expel.

When e're you plant, through oaks your beech diffuse;
The hard male-oak, and lofty *cerrus* chuse.
While *esculus* of the mast-bearing kind,
Chief in ilicean groves we always find.
For it affords a far extending shade ;
Of one of these sometimes a wood is made.
They stand unmov'd, though winter do's assail,
Nor more can winds, or rain, or storms prevail.

To their own race they ever are inclin'd,
And love with their associates to be joyn'd.
When fleets are rigg'd, and we to fight prepare,
They yield us plank, and furnish arms for war.
Fuel to fire, to plowmen plows they give,
To other uses we may them derive.
But nothing must the sacred tree prophane :
Some boughs for garlands from it may be ta'en
For those whose arms their country-men preserve,
Such are the honours which the oaks deserve.

We know not certainly whence first of all
This plant did borrow its original.
Whether on Ladon, or on Maenalus
It grew, if fat Chaonia did produce
It first, but better from our mother-earth,
Than modern rumours we may learn their birth.
When Jupiter the world's foundation laid,
Great earth-born gyants heaven did invade,
And Jove himself, (when these he did subdue.)
His lightning on the factious brethren threw.
Tellus her sons misfortunes do's deplore ;
And while she cherishes the yet-warm gore
Of Rhœcus, from his monstrous body grows
A vaster trunk, and from his breast arose
A hardned oak, his shoulders are the same,
And oak his high exalted head became.

His hundred arms which lately through the air
Were spread, now to as many boughs repair.
A sevenfold bark his now stiff trunk does bind ;
And where the gyant stood, a tree we find.
The earth to Jove straight consecrates this tree,
Appeasing so his injur'd deity ;
Then 'twas that man did the first acorns eat.
Although the honour of this plant be great,
Both for its shade, and that it sacred is ;
Yet when its branches shoot into the skies,
Let them take heed, while with his brandish'd flame,
The Thund'rer rages, shaking Natures frame,
Lest they be blasted by his pow'rful hand,
While tamarisks secure, and mirtles stand.

The other parts of woods I now must sing ;
With beech, and oak, let elm, and linden spring.
Nor may your groves the alder-tree disdain,
Or maple of a double-colour'd grain.
The fruitful pine, which on the mountain stands,
And there at large its noble front expands ;
Thick-shooting hazle, with the quick-beam set,
The pitch-tree, withy, lotus ever wet ;
With well-made trunk here let the cornel grow,
And here Orician *terebinthus* too ;
And warlike ash : But birch and yew repress,
Let pines and firs the highest hills possess :
Brambles and brakes fill up each vacant space
With hurtful thorns ; in your fields walnuts place
And hoary junipers, with chesnuts good
With hoops to barrel up Lyaeus blood.

The difference which in planting each is found,
Now learn ; since th' elm with happy verdure's crown'd :
Since its thick branches do themselves extend,
And a fair bark do's the tall trunk commend ;
With rows of elm your garden or your field
May be adorn'd, and the sun's heat repell'd.
They best the borders of your walks compose ;
Their comely green still ornamental shows.
On a large flat continued ranks may rise,
Whose length will tire our feet, and bound our eyes.
The gardens thus of Fountain-bleau are grac'd,

By spreading elms, which on each side are plac'd.
Where endless walks the pleas'd spectator views,
And ev'ry turn the verdant scene renews.

The sage Corycian thus his native field,
Near swift Oebalian Galesus till'd.
A thousand ways of planting elms he found ;
With them he would sometimes inclose his ground :
Oft in directer lines to plant he chose ;
From one vast tree a num'rous offspring rose.
Each younger plant with its old parent vies,
And from its trunk like branches still arise.
They hurt each other if too near they grow ;
Therefore to all a proper space allow.

The Thracian bard a pleasing elm-tree chose,
Nor thought it was below him to repose
Beneath its shade, when he from hell return'd,
And for twice-lost Eurydice so mourn'd.
Hard by cool Hebrus Rhodop' does aspire ;
The artist, here, no sooner touch'd his lyre,
But from the shade the spreading boughs drew near,
And the thick trees a sudden wood appear.
Holm, withy, cypress, plane trees thither prest :
The prouder elm advanc'd before the rest :
And shewing him his wife, the vine, advis'd,
That nuptial rites were not to be despis'd.
But he the counsel scorn'd, and by his hate
Of wedlock, and the sex, incurr'd his fate.

High shooting linden next exacts your care ;
With grateful shades to those who take the air.
When these you plant, you still should bear in mind
Philemon and chaste Baucis : these were joyn'd
In a poor cottage, by their pious love,
Whose sacred ties did no less lasting prove,
Than life it self. They Jove once entertain'd,
And by their kindness so much on him gain'd ;
That, being worn by time's devouring rage,
He chang'd to trees their weak and useless age.
Though now transform'd, they male and female are ;
Nor did their change ought of their sex impair.
Their timber chiefly is for turners good ;

They soon shoot up, and rise into a wood.

Respect is likewise to the maple due,
Whose leaves, both in their figure, and their hue,
Are like the linden ; but it rudely grows,
And horrid wrinkles all its trunk inclose.

The pine, which spreads it self in every part,
And from each side large branches does impart,
Adds not the least perfection to your groves ;
Nothing the glory of its leaf removes.
A noble verdure ever it retains,
And o'er the humbler plants it proudly reigns.
To the gods' mother dear ; for Cybele
Turn'd her beloved Atys to this tree.
On one of these, vain-glorious Marsyas died,
And paid his skin to Phœbus for his pride.
A way of boring holes in box he found,
And with his artful fingers chang'd the sound.
Glad of himself, and thirsty after praise,
On his shrill box he to the shepherds plays.
With thee, Apollo, next he will contend ;
From thee all charms of musick do descend.
But the bold piper soon receiv'd his doom ;
(Who strive with heaven never overcome.)
A strong-made nut their apples fortifies,
Against the storms which threaten from the skies.
The trees are hardy, as the fruits they bear,
And where rough winds the rugged mountains tears
There flourish best ; the lower vales they dread,
And languish if they have not room to spread.

Hazle dispers'd in any place will live :
In stony grounds wild ash, and cornel thrive ;
In more abrupt recesses these we find,
Spontaneously expos'd to rain and wind.

Alder, and withy, chearful streams frequent,
And are the rivers only ornament.
If ancient fables are to be believ'd,
These were associates heretofore, and liv'd
On fishy rivers, in a little boat,
And with their nets their painful living got.

The festival approach'd ; with one consent
All on the rites of Pales are intent :
While these unmindful of the holy-day,
Their nets to dry upon the shore display.
But vengeance soon th' offenders overtook,
Persisting still to labour in the brook.
The angry goddess fix'd them to the shore,
And for their fault doom'd them to work no more.
Thus to eternal idleness condemn'd ;
They felt the weight of heaven, when contemn'd.
The moisture of those streams by which they stand,
Endues them both with power to expand
Their leaves abroad ; leaves, which from guilt look pale ;
In which the never-ceasing frogs bewail.

Let lofty hills, and each declining ground,
(For there they flourish) with tall firs abound.
Layers of these cut from some ancient grove,
And buried deep in mould, in time will move
Young shoots above the earth, which soon disdain
The southern blasts, and launch into the main.

But in more even fields the ash delights,
Where a good soil the gen'rous plant invites.
For from an ash, which Pelion once did bear,
Divine Achilles took that happy spear,
Which Hector kill'd ; and in their champion's fate
Involv'd the ruin of the Trojan state.
The gods were kind to let brave Hector die
By arms, as noble as his enemy.
Ash, like the stubborn heroe in his end,
Always resolves rather to break than bend.

Some tears are due to the Heliades ;
Those many which they shed deserve no less.
Griev'd for their brother's death, in woods they range,
And worn with sorrow, into poplars change.
By which their grief was rend'red more divine,
While all their tears in precious amber shine.
These, with your other plants, still propagate :
'Tis true indeed they are appropriate
To Italy alone, and near the Po,
Who gave them their first being, best they grow.

Into your forests shady poplars bring,
Which from their seed with equal vigour spring.
Rich groves of ebony let India show ;
Judæa balsoms which in Gilead flow :
Persia from trees her silken fleeces comb ;
Arabia furnish the Sabæan gum ;
Whose odours sweetness to our temples lend,
And at the altar with our pray'rs ascend :
Yet I the groves of France do more admire,
Which now on meads, and now on hills aspire.
I not the wood-nymph, nor the Pontick pine
Esteem, which boasts the splendor of its line ;
Or those which old Lycæum did adorn ;
Or box on the Cytorian mountain born :
Th' Idæan vale, or Erymanthian grove,
In me no reverence, no horror move ;
Since I no trees can find so large, so tall,
As those which fill the shady woods of Gaul.

When from the cloven bud young boughs proceed,
And the mast-bearing trees their leaves do spread ;
The pestilential air oft vitiates
The seasons of the year, and this creates
Whole swarms of vermin, which the leaves assail,
And on the woods in num'rous armies fall.
Creatures in different shapes together joyn'd,
The horrid eruc's, palmer-worm design'd
With its pestif'rous odours to annoy
Your plants, and their young off-spring to destroy.
Remember then to take these plagues away,
Lest they break out in the first show'rs of May.

From planting new, and lopping aged trees,
The prudent ancients bid us never cease :
Thus no decay is in our forests known ;
But in their honour we preserve our own.
Thus in your fields a sudden race will rise,
Which in your nurseries will yield supplies ;
That may again some drooping grove renew :
For trees, like men, have their successions too.
Their solid bodies worms and age impair,
And the vast oak gives place to his next heir.
While such designs employ your vacant hours,

As ordering your woods and shady bow'rs ;
Despise not humbler plants, for they no less
Than trees, your gardens beauty do increase.
With what content we look on myrtle groves !
On verdant laurels ! there's no man but loves
To find his limon, with *acanthus*, thrive.
To see the lovely *philyrea* live ;
With oleander.   Ah ! to what delights
Shorn cypress, and sweet jessamine invites.

If any plain be near your garden found,
With cypress, or with horn-beam, hedge it round.
Which in a thousand mazes will conspire,
And to recesses unperceiv'd retire.
Its branches, like a wall, the paths divide ;
Affording a fresh scene on every side.
'Tis true, that it was honour'd heretofore ;
But order quickly made it valued more,
By its shorn leaves, and those delights which rose
From the distinguish'd forms in which it grows,
To some cool arbor, by the ways deceit,
Allur'd, we haste, or some oblique retreat :
Where underneath its umbrage we may meet
With sure defence against the raging heat.

Though cypresses contiguous well appear ;
They better shew if planted not so near.
And since to any shape, with ease, they yield,
What bound's more proper to divide a field ?
Repine not, *cyparissus*, then in vain ;
For by your change you glory did obtain.

Sylvanus and this boy with equal fire
Did heretofore a lovely hart admire ;
While in the cooler pastures once it fed,
An arrow shot at random, struck it dead.
But when the youth the dying beast had found,
And knew himself the author of the wound,
With never-ceasing sorrow he laments,
And on his breast his grief and anger vents.
Sylvanus mov'd with the poor creature's fate,
Converts his former love to present hate.
And no more pity in his angry words,

Than to himself th' afflicted youth affords.
Weary of life, and quite opprest with woe,
Upon the ground his tears in channels flow :
Which having water'd the productive earth,
The cypress first from thence deriv'd its birth,
With Sylvan's aid ; nor was it only meant
T' express our sorrow, but for ornament.
Chiefly when growing low your fields they bound,
Or when your gardens avenues are crown'd
With their long rows ; sometimes it serves to hide
Some trench declining on the other side.
Th' unequal branches always keep that green,
Of which its leaves are ne're divested seen.
Tho shook with storms, yet it unmov'd remains,
And by its trial greater glory gains.

Let *philyrea* on your walls be plac'd,
Either with wyre, or slender twigs made fast,
Its brighter leaf with proudest arras vies,
And lends a pleasing object to our eyes.
Then let it freely on your walls ascend,
And there its native tapistry extend.
Nor knows he well to make his garden shine
With all delights, who fragrant jassemine
Neglects to cherish, wherein heretofore
Industrious bees laid up their precious store.
Unless with poles you fix it to the wall,
Its own deceitful trunk will quickly fall.
These shrubs, like wanton ivy, still mount high.
But wanting strength on other props relie.
The pliant branches which they always bear,
Make them with ease to any thing adhere.
The pleasing odors which their flow'rs expire,
Make the young nymphs and matrons them desire,
Those to adorn themselves withal ; but these
To grace the altars of the deities.

With foreign jassemine be also stor'd,
Such as Iberian valleys do afford :
Those which we borrow from the Portuguese ;
With them which from the Indies o're the seas
We fetch by ship ; in each of which we find
A difference of colour, and of kind.

Though gentle Zephyrus propitious proves,
And welcome spring the rigid cold removes ;
Haste not too soon this tender plant t' expose.
Your gardens glory, the rash primrose, shows
Delay is better ; since they oft are lost,
By venturing too much into the frost.
The cruel blasts which come from the north wind,
To over-hasty flow'rs are still unkind.
Let others' ills create this good in you,
Without deliberation nothing do.
For this will scarce the open air endure,
Till by sufficient warmth it is secure.

No tree your gardens, or your fountains more
Adorns, than what th' Atlantick apples bore.
A deathless beauty crowns its shining leaves,
And to dark groves its flower lustre gives,
Besides the splendour of its golden fruit,
Of which the boughs are never destitute ;
This gen'rous shrub in cases then dispose,
Made of strong oak, these little woods compose ;
Whose gilded fruits, and flow'rs which never fade,
A grace to th' country and your garden add,
Proud of the treasures nature has bestow'd.
When snowy flow'rs the slender branches load,
And straying nymphs to gather them prepare,
Molest them not, but let your wife be there ;
Your children, all your family employ,
That so your house its orders may enjoy :
That with sweet garlands all may shade their brows ;
For in their flow'rs these plants their vigor lose.
Suffer the nymphs to crop luxuriant trees,
And with their fragrant wreaths themselves to please,
Such soft delights they love ; then let them still
With their fresh-gather'd fruit their bosoms fill.
These apples Atalanta once betray'd ;
They, and not love, o'recame the cruel maid.
These were the golden balls which slack'd her pace,
And made her lose the honour of the race.

But these sweet smells and pleasant shades will cease,
Nor longer be your gardens happiness ;
Unless the hostile winter be represt,

And those strong blasts sent from the stormy east.
Wherefore to hinder these from doing harm,
You must your trees with walls defensive arm.
To such warm seats they ever are inclin'd,
Where they avoid the fury of the wind.
These plants besides that they this cold would shun,
Look for th' Assyrian, and the Median sun.
In parched Africa they flourish more,
Than if they grow by Strymon's icy shore.
Lest then the frost, or barb'rous north should blast
Your flow'rs, while all the sky is over-cast
With duskish clouds, sheds set apart prepare,
To guard them from the winter's piercing air :
Till the kind sun these tempests do's disperse,
And with his influence chears the universe.
Then calmer breezes shall o're storms prevail,
And your fresh groves shall sweet perfumes exhale.

These trees are various, and the fruits they bear,
Are diff'rent too.   The limons always are
Of oval figure, underneath whose rind
A juice ungrateful to our taste we find.
But though at first our palates it displease,
Yet better with our stomach it agrees.
Others less sharp do in Hetruria spring ;
Some, that are mild, from Portugal we bring.
Another sort from old Aurantia came,
To which that city does impart its name.
Hard by Dircaean Aracynthus lies
This ancient town ; the orange hence does rise,
To which in rind and juice the limons yield,
By each new soil new tastes are oft instill'd.

Mind not the fables by the Grecians told
Of the Hesperian sisters, who of old
On vast Mount Atlas, near the Libyan sea,
With greatest care did cultivate this tree
Of fierce Alcides, who by force brake in,
And in the spoils of the Nemean skin ;
And from the dragon, who securely slept,
Stole, with success, the apples which he kept.
Return'd to the Aventine, he sets that hill,
With orange-trees, which Italy now fill.

But things of greater moment are behind ;
For purple oleander may be joyn'd
With oranges, and myrtles ; each of these
Peculiar graces of their own possess.
The myrtle chiefly, which, if fame says true,
From the god's bounty its beginning drew.

When Venus plac'd it in the pleasant shade
Of the Idaean vales, about it playd
Whole troops of wanton cupids, while the night
Was clear, and Cynthia did display her light.
This Citherea above all prefers,
And by transcendent favour made it hers.
With myrtle, hence, the wedded pair delights
To crown their brows at hymenaeal rites.
Hence Juno, who at marriages presides,
For nuptial torches always these provides.
Eriphyle, sad Procris, Phaedra too,
And all those fools, who in Elysium wooe,
Honour this plant, and under myrtle groves,
If after death they last, recount their loves.

Proud victors with its boughs themselves adorn,
While round their temples wreaths with it are worn.
Tudertus, when the vanquish'd Sabines fled,
Plac'd one of these on his triumphant head.
The trunk is humble, and the top as low,
On which soft leaves and curled branches grow.
Its grateful smell, and beauty so exact,
Th' admiring nymphs from ev'ry part attract.
If too much heat, or sudden cold surprize,
Which are alike the myrtles enemies,
You must avoid them both, and quickly place
The tender plant within a wooden-case.
Sheds may protect them, if the cold be great,
Or watring from the summer's scorching heat ;
No impious tool our tenderness allows
To fell these groves, nor cattel here must browse.

Oft oleanders in great vasa's live,
With myrtles mix'd, and oranges, and give
Some graces to your garden, which arise
From the confusion of their diff'rent dies.

In watry vales, where pleasant fountains flow,
Their fragrant berries, lovely bay-trees show,
With leaves for ever green, nor can we guess
By their endowments their extraction less.
The charming nymph liv'd by clear Peneus side,
And might to Jove himself have been ally'd,
But that she chose in virtues paths to tread,
And thought a god unworthy of her bed.
Phoebus, whose darts of late successful prov'd
In Python's death, expected to be lov'd ;
And had she not withstood blind Cupid's pow'r,
The fiery steeds and heav'n had been her dow'r :
But she by her refusal more obtain'd,
And losing him, immortal honour gain'd,
Cherish'd by thee, Apollo.   Temples wear
The bays, and ev'ry clam'rous theater.
The Capitol it self, and the proud gate
Of great Tarpeian Jove they celebrate.
Into the Delphick rites, the stars they dive,
And all the hidden laws of fate perceive.
They in the field (where death and danger's found,
Where clashing arms, and louder trumpets sound)
Incite true courage : Hence the bays, each muse,
Th' inspiring god, and all good poets chuse.

Persian *Ligustrum* grows among the rest,
Whose azure flowers imitate the crest
Of an exotick fowl ; they first appear
When the warm sun and kinder spring draws near,
Then the green leaves upon the boughs depend,
And sweet perfumes into the air ascend.

Pomegranates next their glory vindicate,
Their boughs in gardens pleasing charms create :
Nothing their flaming purple can exceed,
From the green leaf the golden flow'rs proceed :
Whose splendor, and the various curls they yield,
Add more than usual beauty to the field.
As soon as e're the flowers fade away,
Yet to preserve their lustre from decay,
To them the fruit succeeds, which in a round
Conforms it self, whose top is ever crown'd
In seats apart, stain'd with the Tyrian dye,

A thousand seeds within in order lye.
Thus, when industrious bees do undertake
To raise a waxen empire, first they make
Rooms for their honey in divided rows ;
And last of all, on twigs the combs dispose.
So ev'ry seed a narrow cell contains,
Made of hard skin, which all the frame sustains.
Neither too sharp or sweet the seeds incline
Too much, but in one mixture both conjoin.

From whence this crown, this tincture is deriv'd,
We now relate ; the nymph in Africk liv'd :
Descended from the old Numidians race,
Beauty enough adorn'd her swarthy face ;
As much as that tann'd nation can admit,
Too much, unless her stars had equall'd it.
Mov'd by ambition, she desir'd to know
What e're the priests or oracles could show
Of things to come.   A kingdom they dispense
In words including an ambiguous sense.
She thought a crown no less had signify'd,
But in the priests she did in vain confide.
When Bacchus th' author of the fruitful vine
From India came, her for his concubine
He takes ; and to repair her honour lost,
Presents her with a crown ; by fate thus crost,
The too ambitious virgin ceas'd to be ;
Transmitting her own beauty to this tree.

Sharp *paliurus*, *ramnus*, (which by some
Is white-thorn term'd) your garden will become.
There leavy *caprifoil*, *alcaea* too,
Th' Idaean bush, and *halimus* may grow.
Woody *acanthus*, *ruscus* there may spring,
With other shrubs, these skilful gard'ners bring
Into a thousand forms ; but 'tis not fit
To tell their species almost infinite.

From brighter woods the prospect may descend
Into your garden, there it self extend
In spacious walks, divided equally,
Where the same angles in all parts agree.
In oblique windings others plant their groves,

For ev'ry man a diff'rent figure loves.
Thus the same paths, respecting still their bound
In various tracts diffuse themselves around.
Whether your walks are straight or crooked made,
Let gravel, or green turf be on them laid.
The nymph and matrons then in woods may meet,
There walk, and to refresh their weary'd feet,
Into their chariots mount, tho' to the young
Labour and exercise does more belong.

If close-shorn *philyrea* you deduce
Into a hedge, for knots the carpine use ;
Or into arbors with a hollow bark,
The pliant twigs of soft *acanthus* make.
With stronger wires the flowing branches bind,
For if the boughs by nothing are confin'd,
The tonsile hedge no longer will excell ;
But uncontrol'd beyond its limits swell.
And since the lawless grass will oft invade
The neighb'ring walks, repress th' aspiring blade,
Suffer no grass or rugged dirt t' impair
Your smoother paths ; but to the gardners care
These things we leave ; they are his business,
With setting flow'rs, and planting fruitful trees :
And with the master let the servants join,
With him their willing hearts and hands combine :
Some should with rowlers tame the yielding ground,
Making it plain where ruder clods abound.
Some may fit moisture to your medows give,
And to the plants and garden may derive
Refreshing streams ; let others sweep away
The fallen leaves ; mend hedges that decay ;
Cut off superfluous boughs ; or with a spade
Find where the moles their winding nests have made ;
Then close them up : Another flow'rs may sow
In beds prepar'd ; on all some task bestow ;
That if the master happens to come down,
To fly the smoak and clamour of the town ;
He in his villa none may idle find,
But secret joys may please his wearied mind.

And blest is he, who tir'd with his affairs, (
Far from all noise, all vain applause, prepares

To go, and underneath some silent shade,
Which neither cares nor anxious thoughts invade,
Does, for a while, himself alone possess ;
Changing the town for rural happiness.
He, when the sun's hot steeds to th' ocean hast,
E're sable night the world has over-cast,
May from the hills the fields below descry,
At once diverting both his mind and eye.
Or if he please, into the woods may stray,
Listen to th' birds, which sing at break of day ;
Or, when the cattle come from pasture, hear
The bellowing oxe the hollow valleys tear
With his hoarse voice : Sometimes his flow'rs invite ;
The fountains too are worthy of his sight.
To ev'ry part he may his care extend,
And these delights all others so transcend,
That we the city now no more respect,
Or the vain honours of the court affect :
But to cool streams, to aged groves retire,
And th' unmix'd pleasures of the fields desire ;
Making our beds upon the grassie bank,
For which no art, but nature we must thank.
No marble pillars, no proud pavements there,
No galleries, or fretted roofs appear,
The modest rooms to India nothing owe ;
Nor gold, nor ivory, nor arras know :
Thus liv'd our ancestors when Saturn reign'd,
While the first oracles in oaks remain'd :
A harmless course of life they did pursue ;
And nought beyond their hills, their rivers knew.
Rome had not yet the universe ingross'd,
Her Seven Hills few triumphs then could boast.
Small herds then graz'd in the Laurentine mead ;
Nor many more th'Arician valleys feed.

Of rural ornaments, of woods much more
I could relate, than what I have before ;
But what's unfinish'd, my next care requires,
And my tir'd bark the neighb'ring port desires.

Resonate Montes Laudationem, SILVA,
Et omne Lignum *ejus.* Isa. 44. 23.

Printed in the United States
By Bookmasters